VIGNETTES

Vignettes

Series Editors: Tony Hughes-d'Aeth and Sarah Collins

The Vignettes Series from UWAP is aimed at sharing the knowledges that are emerging in the contemporary university and that consider the complexities of modern life. Each book provides an image, or a vignette, of a particular phenomenon and how this is being thought through by intellectual practitioners in today's academy.

Books in This Series

- *Netflicks: Conceptual Television in the Streaming Era* by Tony Hughes-d'Aeth
- *The Moment of the Essay* by Daniel Juckes
- *Saving Heritage Breeds: a Love Story* by Catie Gressier
- *Artificial Life* by Oron Catts, Sarah Collins, Elizabeth Stephens and Ionat Zurr (forthcoming)
- *Australiana* by Graham Akhurst and Penni Russon (forthcoming)

VIGNETTES

Saving Heritage Breeds

A Love Story

CATIE GRESSIER

First published in 2025 by
UWA Publishing
Crawley, Western Australia 6009
www.uwap.uwa.edu.au

UWAP is an imprint of UWA Publishing
a division of The University of Western Australia

UWA Publishing acknowledges we are situated on Noongar land, and that Noongar people remain the spiritual and cultural custodians of their land, and continue to practise their values, languages, beliefs and knowledge. We pay our respects to the traditional owners of the lands on which we live and work across Western Australia and Australia.

ISBN: 978-1-76080-302-5

A catalogue record for this book is available from the National Library of Australia

Cover design by Mika Tabata
Typeset in 11 point Garamond by Lasertype
Printed by McPherson's Printing Group

For Abbie

Contents

Introduction

Greg was the youngest in his family and always wanted to farm. But there were six kids, and not enough land to support them all, so his parents encouraged him to get a trade. 'But I just loved it', he told me:

> As a child, I had pine cones as my herd of cows. And I used to go and get out there, and cut oats with Mum's scissors, you see, to go feed my 'cows'. So that's all I wanted to be.

Greg's childhood wasn't all idyllic, though; farms can be dangerous places. When he was little, he fell into a large, stainless-steel bucket of boiling water resulting in third-degree burns to more than 30% of his body. He choked up when he described still having his dad's handprint on his chest from where he had pulled him out of the boiling water. His doctors didn't think he'd survive, and he and his mum lived for four months in the Melbourne Children's Hospital, while his dad did the milking and looked after the other kids.

Greg's middle brother also diced with death as a child, when he was pinned and gored by a Jersey bull. He still wears the scars of that bull's horns across his stomach.

Greg's daughter, Brittany, had her own run-in with a Jersey bull a couple of years ago when she was working in an artificial insemination centre not far from their farm on Girai wurrung Country in Western Victoria. Despite the traumatic experience of being rammed repeatedly by an enormous, irate bull, her love for cattle remains strong, and she is training to be a livestock vet. She was taught to handle cattle by her grandmother, who Greg describes as 'a real cow woman', like her dairying mother before her. When Greg's mum met and married Greg's dad—a skilled cattleman in his own right—his grandmother gave the newlyweds two of her best cows, two buckets and a separator (to separate the cream from the skim milk) to get them started off in their own dairy.

After school, Greg worked for twenty years as a plumber, but never relinquished his plans to farm. Recognising Greg's determination, when his father was dying from leukaemia, he gave Greg his Red Poll cattle. The earliest recorded Red Polls in Australia were the herd belonging to the Reverend Samuel Marsden in 1808. Red Polls soon became among the most popular house cows, being a dual-purpose animal that could provide a large volume of milk with high butterfat content, and quality beef with good marbling. They were valued for being low maintenance and hardy, thriving in marginal conditions, while being easy to handle and having good temperament.

From mid last century, however, the industrialisation of agriculture saw mixed-enterprise family farms being bought out by large, targeted agribusinesses. Profits were maximised through selectively breeding for a singular purpose, with cattle divided into beef or dairy breeds. The number of dual-purpose Red Polls has since dropped dramatically, which made perpetuating the bloodline his father had been developing for thirty years all the more important for Greg. He described to me his early years of farming:

> We were in a drought when I started leasing ground, and I just didn't want to sell this beautiful line of heifers. So, the first year of rent was the hardest. I remember selling cull cows to pay for the rent. We had a young family, you know, young kids under ten; they were tough times. So, we crawled, and scraped, and baled hay off the side of the bloody road to survive during that dry year…
>
> I had to learn the hard way, and one way to survive is having the right type of cow. You know, cows that have a moderate frame, and not a big-framed animal, which takes a lot to maintain. So, the feed efficiency of Red Polls! How can you compare that to the cow next door? Because people just think, you know, a cow's a cow, but it comes down to, yeah, they just don't let you down.

Red Polls are listed as endangered by the Rare Breeds Trust of Australia (RBTA). Yet, increasing interest in the breed's qualities, particularly their excellent maternal characteristics and temperament, hopefully signals a brighter future for this wonderful old breed.

Livestock in Australia

Even while Red Polls are considered an 'old' breed, livestock, of course, is relatively new to Australia. In the late eighteenth century, over 65,000 years of First Nations peoples' connection to Country was violently disrupted by British colonial occupation. In most colonies, globally, settlers incorporated native foods into their diets. Yet, British settlers' sense of cultural superiority extended to dietary practices. This meant that the colony's focus was on growing food using animals and crops from Europe. Farming has consequently played a substantial role in the ongoing dispossession of First Nations Australians. Agricultural productivity provided a sense of legitimacy and belonging for settlers, while violent massacres resulted from contestation over land and water for grazing, or retribution for First Peoples' hunting of livestock. My own family is implicated in this ugly history. Arriving as free settlers to the colony in 1805, my family has farmed cattle and sheep in New South Wales ever since.

Over time, many First Nations people have become integrally involved in livestock production. This was initially as coerced, unpaid labour; yet, many First Peoples have come to be highly skilled and passionate pastoralists. First Nations groups have also, in some instances, incorporated introduced animals not only into their dietary practices and vocations but into their cosmologies too.

Livestock-keeping practices have changed significantly since the early years of the colony. Mechanisation, genetic technologies, and the use of fertilisers, herbicides and veterinary pharmaceuticals have enabled the unprecedented scaling up and intensification of animal production. This has vastly increased yield and efficiencies, but has compromised animal welfare and led to ecological degradation and biodiversity loss. Intensification has also resulted in growing zoonotic disease prevalence and increased antimicrobial resistance. The 'get big or get out' approach has altered the size and composition of rural communities. Small-scale Australian farmers today struggle to make a living in the face of global competition, a duopolistic domestic retail market, the changing and unpredictable climate, an increasingly complex and bureaucratic regulatory environment, and dynamic consumer values. Agricultural services, most notably abattoirs and processing facilities, are increasingly centralised and difficult for small-scale producers to access.

All of these issues are widely understood. Much less well known are the biodiversity losses occurring within domestic species. It has taken thousands of years of natural and artificial selection for the extraordinary diversity of domestic animals to develop. Yet, in less than a century, hundreds of unique cattle, sheep, pig and poultry breeds have become extinct. This has occurred largely as the result of multinational agribusinesses investing heavily in breeding programs and marketing campaigns that promote and widely disseminate the genetics of an ever-diminishing number of hyper-productive and profitable breeds and bloodlines.

When I learned about the extinction crisis unfolding on our farms, I was shocked that more people weren't talking about it. Then again, while livestock continues to dominate the biosphere at a cost to ecosystems and wildlife, it is fair to question whether livestock breed extinctions matter. While less meat in our diets would improve ecological and animal welfare outcomes, the reality is that animal consumption is deeply entrenched in Australians' dietary practices and the nation's economy. Consequently, supporting ecologically sustainable and high-welfare approaches to livestock production is critical. More broadly, manifestations of climate change in the form of the catastrophic droughts, fires and floods of recent years have made the urgency of learning to live less destructively, and in ways respectful to all life forms, abundantly clear.

Resilient systems need to be able to tolerate stresses and maintain value and capacity, while not undermining the natural resource base into the future. Scientists have long understood that risk is mitigated by ensuring diversity at the genetic, species and ecosystem levels. Livestock breed diversity is part of this picture and provides resilience given the many unpredictable challenges ahead, including our changing climate, emergent disease risks and shifting market demands.

Across Australia, a passionate subset of farmers is working hard to maintain agrobiodiversity through the preservation of heritage breeds, whose bloodlines and histories are enmeshed with their own. Unlike in Europe, where agrobiodiversity is valued and government subsidies reward the conservation efforts of heritage breed farmers, there are no government incentives for raising heritage breeds in Australia. Consequently, creating a commercial demand for meat, milk, eggs and fibre is the dominant means of ensuring breed conservation. Yet, heritage breed farmers face substantial challenges in doing so. Economic viability is hampered by consumers' reluctance to pay more to cover the costs of raising slower growing breeds in free-range environments. Small genetic pools render inbreeding a perennial risk, while climatic events, such as drought, flood and fire, threaten localised populations.

So why do these farmers persevere with heritage breeds? Why not make life easier and make more money with commercial breeds and hybrids? Over four years I conducted research with heritage breed farmers across Australia and asked these questions repeatedly. Irrespective of which breeds they raise, the answer farmers give time and again is, quite simply, that they love them. This book tells the story of heritage breed cattle, sheep, pigs and poultry, and lays bare the role of love in breed conservation. Farmers are consummate storytellers, and I have adopted their approach and crafted this book as a series of anecdotes interspersed with some explanations of the history of the core livestock species that we eat in Australia. You will meet cattle, sheep, pig and poultry farmers from across the country and gain a window into their lives and those of their beloved animals.

Agrobiodiversity

But first, how did we end up in this situation? Over 40% of the world's daily calories are provided by just three crops: rice, wheat and maize. Increasing reproductive problems in the now narrow genetic pool of microbes used by commercial cheesemakers to make camembert and roquefort is resulting in the threat of extinction of some of the world's favoured *fromages*. And while there are over a thousand known varieties of banana, global trade is now dominated by just one: the cavendish. These are just a few examples of the homogenisation of our diets globally.

The variety in our diets has shrunk hand in hand with globalisation, trade liberalisation and the rise of multinational food corporations. Westernisation of global dietary practices has seen the preferencing of animal products, and ultra-processed foods rich in salt, sugar and plant oils, at the expense of traditional vegetables, cereals and pulses. Human health has been negatively impacted. Overnutrition has overtaken undernutrition in many parts of the world, and non-communicable diseases, including diabetes, heart disease and certain forms of cancer, have risen in prevalence.

Small mixed-crop and livestock farms have been subsumed by large agribusinesses that produce this shrinking variety of crops and livestock for global commodity markets. Control of our food system has moved out of the hands of communities and into highly organised and capitalised corporate chains. In Australia, our diets are largely determined by two major supermarkets. Their meat offerings are all about scale, uniformity and consistency, which has proved very bad news for heritage breeds. Butchers have been majorly disenfranchised by the supermarkets' dominance, and the anger and frustration of one Perth-based butcher is clear:

> I understand that rare breeds need to be saved if we're wanting meat for the future, if we're wanting diversity. But Coles and Woolworths and Aldi, they own 90 cents, if not 95, of every retail dollar. They took over meat. They destroyed 90% of the butcher shops. They took over fruit and veg. They destroyed 90% of fruit and veg shops. They will eventually destroy all the small guys. I'm a fucking kid who left school at fourteen, and I can see it! Can't government see? The farms are being absolutely brutalised.

As a major food exporter, complacency around food security prevails in Australia. Yet, producing vast amounts of food does not necessarily ensure a food-secure future if production is ecologically unsustainable, and resources are not wisely managed.

Within this broader pattern of dietary homogenisation, the Food and Agriculture Organization (FAO) confirms that at least one in ten livestock breeds is already extinct. The loss is accelerating, with over 25% of all livestock breeds at serious risk of extinction. Many of the wild species from which these domestic animals originated are also extinct. The problem is particularly pronounced in wealthier nations, where the dissemination of the most profitable breeds and bloodlines has been widespread. More resource-constrained nations continue to be the protectors of agrobiodiversity, as farmers work in environments where local breeds are valued for their pest and pathogen resistance, hardiness and suitability to local conditions, both ecologically and culturally.

Of further concern is that genetic diversity within breeds, both heritage and commercial, is also diminishing. Conservation biology recognises the viability of a population as directly linked to its effective population size. If this reaches below fifty, this seriously compromises the survival of a wild population. Not only heritage breeds, but also many commercial breeds who favour a small number of hyper-productive bloodlines, now suffer from inbreeding, with low effective population sizes. Inbreeding enhances both positive productive traits, but also negative ones. Loss of fertility and increasing disease prevalence have emerged in inbred populations over the past decades.

In Australia, the loss of agrobiodiversity is pronounced. All commercial chicken meat in Australia derives from just two types: hybrid strains of the Ross and Cobb, whose genetics are owned by two multinational agribusinesses. Dairy cattle fare little better. The normalisation of cheap milk has seen countless family dairy farms close, as the economies of scale required to meet market demands favour large-scale operations. High-output Holsteins supply over 70% of Australia's milk, while Angus dominate in the beef industry. Eleven Australian cattle breeds are already extinct, with a further twenty-two breeds endangered. Similarly, Merinos produce more than 80% of the nation's fleece, while fourteen sheep species are critically endangered owing to the prime lamb market being dominated by Merino hybrids. Pigs are perhaps in the worst shape, with all pig breeds having low numbers of registered purebred animals.

In Australia, there are no cryo-conservation facilities systematically collecting the sperm and embryos of at-risk livestock breeds. Conservation is reliant on live animal populations and is relegated to the private sphere and the efforts of small-scale farmers and volunteer-run rare breed societies, particularly the RBTA.

The loss of agrobiodiversity raises many questions. How do we make sense of the careful breeding over thousands of years of particular animal types, only to allow these forms to become extinct in a matter of decades? Should we even worry about the loss of breed diversity in livestock, which has been implicated in the eradication of many native species globally as they compete for increasingly limited land and water resources? Is it better for a breed to die out, rather than being perpetuated only to have individual animal lives extinguished prematurely for human uses? Or, if the quality of life and high welfare of animals is ensured, is this ample justification for the use of animal products? Can the individual animal be legitimately sacrificed for the perpetuation of the bloodline or breed more broadly? There are no easy answers, but hopefully this book will offer some food for thought on these questions.

What Is a Breed?

But what even is a breed? While breeds today may seem like an evidential fact, the term only emerged in the seventeenth century in France regarding cattle, and around the same time in Germany for horses. While there have long been clearly differentiated livestock types borne of geographic isolation and adaptation to different environments, for about 98% of their roughly 10,000 years of history living alongside people, little selection pressure was applied to domestic animals. Indeed, in many parts of the world, livestock are still left to make their own breeding decisions; they are not fenced and closely managed as they are in Australia today. This point was reinforced for me when I messaged a friend in Botswana to ask what breed of cattle he had. His reply was quick, and still makes me smile: 'Eish! I don't know *where* my cattle are, never mind *what* they are!' Physical, and in turn genetic, isolation of breeds has thus historically seldom been complete.

Selective breeding as we know it was pioneered in England by Robert Bakewell with Longhorn cattle and Leicester sheep, and in Shorthorn cattle by the Colling brothers. Their radical approach entailed the introduction of breeding objectives, the careful selection of breeding individuals, and the denial of reproduction to those not meeting the ideal type. Pedigrees were documented in herd books. Bakewell's business model involved keeping tight control over his improved animals, while forming elite breed societies and selling his genetics at a premium. This constituted a major shift in the industry and reflected a broader societal move towards commoditisation: that is, away from intrinsic and use value, and towards market value.

The breed concept sits at an intersection between science and culture. While definitions remain somewhat contested, many would agree that a breed is a type of domestic animal possessing a particular series of traits and characteristics. This is owing to reproductive isolation, which historically has usually been the result of some level of human control, along with adaptation to a particular environment. Yet, a breed can't be considered a genetic concept so much as a descriptive one, as breeds do not always correlate with distinct genetics. That is, in some cases, even where there are clear differences in appearance, animals from different breeds may differ only slightly more than those from the same breed.

As breeders have long observed, and now genetic research confirms, even within closed herds and flocks that have been inbred to encourage homogenous stock and a consistent phenotype, there is sufficient diversity within the genome to ensure a considerable amount of variation between individuals. Furthermore, distinct breeds in many cases share common characteristics, which some breeders who are less concerned with purity take advantage of. As Grant, a poultry breeder from New South Wales, told me:

> The genetic integrity of fowls is something that I think is very open to question, because so many breeds are composite breeds. So people will still mate this and that and get birds that look like what they're aiming for, and then market them or show them as being that particular breed. Whereas, in fact, they might be a first-generation cross with two similar looking birds. And because there's no stud book as such for fowls, there's more of that sort of thing going on than could possibly go on with cattle, where there's a stud book.

As a consequence of this lack of definitive genetic difference, many definitions emphasise the social construction of breeds. A popular definition to this end is 'a breed is a breed if enough people believe it to be'. Recognising the politics inherent in notions of breed, another tongue-in-cheek definition suggests 'a breed is a breed if the breeders can get along'. Still others suggest that because international conventions on biological diversity relegate control to each nation, that 'a breed is whatever the government claims it to be'.

While it is certainly valuable to preserve breeds given their unique physical characteristics, there is arguably a paradox inherent in keeping groups genetically separate given genetic isolation decreases diversity. Further, the guiding principle of 'improvement' that has underpinned agricultural enterprise over the past few hundred years has both increased and been a profound threat to breed diversity.

This brief discussion doesn't nearly do justice to the history or science of breeds and breeding. But my point is that breeds are a complex entity. One of the few certainties is that breeds are and will remain dynamic. Both commercial and heritage breeds have changed considerably over the decades and will no doubt continue to evolve into the future. Notwithstanding these layers of complexity, the concept of breed is widely agreed to be valuable for managing agrobiodiversity conservation.

Heritage Breeds

You won't be surprised to learn that breeds deemed 'heritage' are similarly subject to differing definitions. In Australia, heritage breeds are generally considered to be purebred animals with a documented history of breeding true to type for at least forty years, who existed prior to the development of today's commercial types. The notion of 'heritage' is, of course, a cultural construct. Interestingly, in Canada, this is made explicit in their definition of a heritage breed that includes the requirement to have been significant within the nation's agricultural history. In the USA, there is an emphasis on heritage breeds as possessing self-sufficiency and survival traits, such as longevity, fertility and foraging ability, which have developed over time to ensure adaptation to local ecosystems. These traits have enabled heritage breeds to thrive in traditional farming systems, where animals have historically been relatively self-sufficient, and not reliant on the high feed and veterinary inputs and carefully controlled environments normalised in intensive systems. A further characteristic of heritage breeds is that not all, but most, are in varying levels of endangerment.

Like the breed concept broadly, there is contestation and critique around the principles and practices associated with heritage breeds. In the USA and Europe, there is more cultural value placed on heritage breeds, so breeders can attract higher prices for their products. This is less established in Australia, where breed is less often used as a marketing tool. Yet, in the US, one of the critiques of heritage breeds is that it is a niche market for the wealthy. By contrast, in Australia, commercial breed farmers are more likely to see heritage breeds as an anachronism or a romantic folly.

There are polarised views around the position of livestock broadly, of course, with some suggesting livestock should be reduced by any means possible given their contribution to global greenhouse gas emissions and ecological decline. There is little sympathy from this camp about preserving breed diversity. Many commercial breed farmers value heterosis, or hybrid vigour, given the progeny of genetically different animals can have positive traits, such as increased fertility and growth characteristics. The concern for preserving, or at least registering, purebred animals is of less concern for some in this camp. Others see the faster growth of commercial breeds and the efficiency of intensive livestock production as the best way to 'feed the world'. They argue that the extensive, free-range practices of most heritage breed farmers do not produce enough meat. Heritage breed proponents would counter this by saying we should all be eating less, but higher welfare, meat that is raised integrated into ecosystems, without the polluting waste produced in intensive systems and the animal welfare issues in situations of confinement.

There is another tension that plays out between heritage breeders who value the preservation of traditional traits and those who wish to develop desirable traits that may constitute a departure from the historical norm. In fact, the irony is that some of the breeds considered heritage today were among those subject to the earliest improvements. Bakewell's development of the English Leicester sheep endangered many other British breeds through their successful dissemination. Yet, today, given the dominance of Merino sheep, English Leicesters have become an endangered heritage breed.

A final complexity relates to the emplaced nature of heritage breeds, which were formed in and adapted to particular ecosystems. Breeds are often named for the regions in which they have lived for a significant amount of time, not solely as geographical indicators, but because they are felt to embody the character of a place. Notions of what is native are problematic, however, given the long histories of mobility for virtually all domestic species. British sheep breeds are named by locality, as one example, and much is invested in notions of their nativeness. Yet, all descend from the mouflon of the eastern Mediterranean.

In Australia, all domestic animals were introduced after colonisation commenced in 1788. Some local producers believe that, rather than fixating on preserving breeds native to elsewhere in the world, we need to work more to develop breeds ecologically adapted to Australian conditions.

So, in short, from these spatial and temporal tangles, through to the tensions around tradition and improvement, there is a complex set of ideas and debates that circulate in the world of heritage breeds.

Heritage Breed Farmers

So, who are heritage breed farmers? To start, it is important to make clear that, in many respects, they are not that different to small-scale farmers with commercial breeds. The latter are often equally passionate about farming and their animals. They face similar challenges and enjoy similar lifestyles, while living in the same communities.

Heritage breed farmers also have considerable internal diversity. Yet, there are two broad categories that emerged in my research. Firstly, quite a lot of those raising heritage breeds are intergenerational farmers whose family has kept particular breeds, often over several generations, and who remain committed to perpetuating their family's chosen breeds. There is also a cohort of first-generation farmers who tend to be passionate about agroecology and regenerative farming, within which heritage breeds are a natural fit. They constitute a distinct subgroup on account of their motivations around food sovereignty and sustainable food production.

Traditionally, farming is conceived as a masculine role, despite the fact that women have always lived and worked on farms. Yet, I have found women to be strongly represented in raising heritage breeds. Across all species, they are centrally involved in heritage breed farming, from day-to-day farm management through to showing animals and participating in breed societies and advocacy.

Farming is often described as a labour of love, but along with this comes a complex set of conditions. Farming is notable for its long working hours in all sorts of weather. It involves diverse, physically demanding tasks. Farmers are unable to readily take holidays owing to animal care responsibilities. Elevated levels of debt and financial risk, alongside irregular income, make for a complex business model. Farmers can struggle with social isolation given their long working hours and the urbanisation of the past half-century. The centralisation and loss of critical infrastructure, particularly abattoirs and meat-processing facilities, adds to farmers' difficulties, as does increasingly complex bureaucratic requirements and regulatory restrictions. High levels of stress are ever more frequently induced by natural disasters and the impacts of climate change.

Intergenerational conflict can be heightened by these factors, along with the lack of separation from the home and work environment. Tammi, a heritage breed pig farmer, described additional common stresses:

> Those constant blows of your animals coming back from the abattoir damaged. Dealing with people who don't want to deal with you, because they want to deal with big industrial types. Planners treating you like you're trying to destroy the environment. Food safety people treating you

> like you're trying to kill everyone. [And this is] when you've got into this because you're trying to save heritage breeds and biodiversity in agriculture, and feed people nutritious food. I think some people's emotional resilience for the constant blowbacks is just tried for too long.

Since 2020, a host of specific challenges have made life even more difficult for farmers. The pandemic interrupted supply chains, closing abattoirs and restaurants, among other critical aspects of heritage breed businesses. Furthermore, several serious diseases have emerged in recent years, with Japanese encephalitis outbreaks and the looming threat of foot-and-mouth disease. Across the country, increasingly frequent and severe drought, bushfires and floods have devastated rural communities. In the chapters that follow, we explore how these issues impact heritage breed farmers.

Researching Heritage Breeds

I'm a cultural anthropologist, which is not the most obvious disciplinary background for a study of heritage breed farming. But much of my research over the years has examined human-nature relations, with a focus on hunting and farming, and foodways more broadly. In 2020, I received an Australian Research Council grant for myself and two PhD students, Tammi Jonas and Domenico Volpicella, to conduct research with heritage breed farmers across Australia.

Since kicking off this project in early 2020, I have driven thousands of kilometres visiting heritage breed farms across every Australian state (though not the two territories). I still laugh when I think of my first farm visit. While no stranger to starting new projects, it always takes a bit of time to find your feet in a new field site (in this case, quite literally a paddock). Kerry had the only flock of Romney sheep in Western Australia. I was super excited to meet this unique flock of sheep and said as much prior to visiting. On arrival, she immediately took me out to her flock. I am not sure quite what she expected from this visiting researcher, but she looked at me expectantly when we reached the animals. As an anthropologist, I had no instruments of measurement to whip out, no samples to take, and despite my grandparents having been sheep farmers, at that point barely knew which end of a sheep was which. So, I just stood there awkwardly patting the thick fleece of an obliging ewe, while yabbering away trying to explain to her (and myself) what exactly it was that I was trying to achieve with this project.

Thanks to the generosity of dozens of farmers like Kerry, I soon found my feet. I undertook participant observation on farms, at agricultural shows, in butchers and restaurants. I spent hours sitting around kitchen tables over cups of tea learning about the rewards and challenges of raising heritage breeds. I recorded interviews with over 60 people, mostly heritage breed farmers, but also farmers who raise commercial breeds, along with many others involved in agriculture and conservation more broadly. While staying on farms, I tried to make myself useful by helping out with the usual chores.

I helped herd cattle, feed poddy calves, draft sheep, split and haul wood, and had my gumboots nibbled by hundreds of excitable piglets. I helped herd errant lambs, fix fences and drench sick sheep and cattle. I clumsily roustabouted during shearing and cooked lunch for farm workers. On one farm, I was relentlessly pecked, day after day, by a broody Brahma while collecting her eggs. I participated in livestock handling workshops and learned how to lead and show cattle. I spent whole days cleaning out freezers, and sorting, packing, labelling and storing meat. I moved mobile chicken pens to fresh grass, fed many a chook and pig, and carted and fed out pellets and hay for sheep and cows. I mucked out urine- and faeces-soaked straw from sheep exhibits, and I donned a high-viz vest and worked crowd control during sheep judging and ram sales at the Perth Royal Show. I hung out with Dexter and Highland cattle breeders while they groomed their cattle for showing. On two separate occasions, I participated in the home kill and processing of an Angus heifer and a Large Black pig. I shadowed judges at poultry shows, hooned around farms on quad bikes, and I even went to my first rodeo. But most enjoyable of all were the many hours spent wandering around paddocks in all kinds of weather with farmers passionately detailing the biographies and lineages of their favourite heritage breed animals.

Structure of the Book

I was drawn to publish within the University of Western Australia Publishing's Vignettes series because this book is exactly that: a series of vignettes. Growing out of the fieldwork described, each chapter offers anecdotes that aim to paint a picture of the lives of various heritage breeds, and the farmers who raise them. The chapters are organised by species. While each has the common threads of heritage breed endangerment and love, and the changes over time for the species, every chapter has a slightly different focus.

We begin with cattle, and an exploration of both the erosion and conservation of biodiversity. We explore loss and hope through heritage breed cattle farmers' experiences, ranging from bushfires through to efforts towards progressing reconciliation with First Nations communities. Contemporary Australia was famously built on the sheep's back, and in the following chapter we explore the role of sheep in First Nations dispossession, and as the foundation of the colonial nation's economy. We learn about the Merino sheep's dominance, and its costs in terms of breed diversity losses. Pigs and poultry follow. The intensification of pork, chicken and egg production has resulted in profound changes to these animals' bodies and lifeways. We explore the radical end of agrarian politics through heritage breed pig farmers and their passionate work in seeking an agroecological transition. Last, but certainly not least, we explore the fascinating history of poultry—mostly chooks—as they've travelled from the jungles of Southeast Asia to the factory farm. We learn about the love heritage chicken people have for their useful and quirky companions, and the fascinating subculture of chicken people, from 'chicken maths' through to poultry shows.

A Love Story

As an anthropologist, I've long been interested in why people do what they do. Often defying economic rationalism, the primary factor motivating heritage breed farmers' conservation efforts seems to be love. But love is something that isn't talked about all that much in farming literature. On the contrary, love gets a bit of a bad rap, as it is often seen as being a bit irrational. However, when it comes to conservation, love acts as a powerful motivating force. Humans save what they love: we know this from wildlife conservation and it holds, too, for livestock. The love heritage breed farmers feel for their animals motivates their ongoing commitment. It is the reason they keep less profitable breeds in the first place. It is the reason they get out of bed every morning and work so hard to care for their animals. It's the reason they spend their weekends at shows promoting their breeds. The emotional distress they experience when they lose favoured animals, or the decisions they make to keep animals for years beyond their productive age at a considerable financial cost, are all underpinned by love.

Significantly, a growing body of scientific evidence confirms that love is good for both animals and people. Animal health, welfare and performance measurably improves with positive animal-handling practices. In turn, heart rate and arterial blood pressure decrease for humans during positive contact with animals, including petting, feeding and talking. Recognising this, using farm animals in therapeutic ways is becoming increasingly popular, and may well be a growing avenue of alternative income for farmers. Yet, for the most part, creating a commercial demand for meat, milk, eggs and fibre remains the only viable means of ensuring heritage breed survival in Australia, paradoxical as this may seem, given it entails killing beloved animals to sustain life. But it's important to emphasise that love does not exclude killing in this context. Love involves making difficult decisions for individual animals with the goal of ensuring good outcomes for the breed.

To wrap up this chapter, I want to return to where we started. My first conversation with Greg about Red Polls took place fairly early on in my research. I had to cut the call short, as I was late for another meeting. I was grateful that Greg was keen to continue the conversation, because as he put it: 'It's a fairly big story. And it depends how far you want to look into it, but it's done with a lot of passion.'

It is a big story, this story of heritage breeds. It's a story that starts with the soil, and the microbes and carbon that are either nourished and sequestered, or leached and released. It's a chronicle of rain and its absence, dry years, drought decades, and fires, floods and disease. It's an account of legacies, heritage and history being negotiated daily with the inevitability of change. It's the story of selecting and perpetuating bloodlines, building up the herd through killing; selling cull cows, and breeding into the herd fertility, docility, and mothering ability. It's a tale of blood, milk, sweat, pathogens and tears—shared across species in high-stakes entanglements of living, loving, dying and killing. It's a story of multispecies lineages: parents handing down their carefully curated genetics, time-tested traits, to their daughters and sons; and bull-wrangling men and women passing on bloodlines and battered bodies, marked forever by bulls' horns, brands and hands. And it's a story that is important to tell, and tell persistently, so its ending isn't one of extinction, of homogenous genes and landscapes, with biodiversity and bloodlines lost forever.

Chapter 1

Cattle

Over the past decade, cattle have become the poster child of agriculture's contribution to global warming. Cattle belch climate change inducing methane into the atmosphere. We watch the ongoing deforestation of the Amazon with horror as cattle ranching encroaches to feed our ever expanding, protein-coveting human population. These negative impacts are an important part of the cattle story. But over 10,000 years of domestication, these remarkable animals—who exhibit greater breed diversity than any other livestock species—have also been of extraordinary ritual, symbolic and economic significance to diverse cultural groups on every inhabited continent. More nuance is needed in public understandings of the impacts of cattle.

Animal agriculture is a contributor to climate change and is also highly exposed to its effects. Farmers and their animals suffer some of the worst exposure to the increasingly frequent droughts, fires and floods of the climate change era. On February 6, 2022, a raging bushfire swept through WA's Wheatbelt and across the farm of Murray, a Red Poll cattle breeder living on Ballardong Noongar Country. Murray, in his seventies, is a highly respected breeder, cattle judge and long-term member of the Australian Red Poll Cattle Breeders Inc. When the fire hit, Murray was 70 kilometres away at his sister's house recovering from a knee replacement.

That summer was the hottest on record—though that record has been broken again since—and stubble from the season's bumper wheat harvest provided plenty of fuel. As the temperature hit 42.5 degrees, wind gusts of over 70 kilometres an hour fanned the flames. Despite the searing heat and suffocating smoke, Murray's 18-year-old farm manager, Ollie, managed to move all the various mobs of cattle out of the fire's path. In the meantime, the fire destroyed several sheds, as well as the house that Ollie was living in. He lost all his worldly possessions that day while saving the cattle.

Bushfire events, along with floods and disease outbreaks, are a constant threat for farmers. But they are particularly concerning in the case of endangered heritage breeds. In Murray's case, his bloodlines reflect decades of careful breeding, with a focus on excellent udders and teats to ensure milking ability, along with growth and muscling traits. The loss of his much-loved herd would have been hard to bear. Over 45,000 hectares were damaged in this fire, with thousands of sheep killed, and a heavy toll on the community, wildlife, vegetation and property. While the impacts of this disaster will be felt for decades to come, there were some positive outcomes.

I had spent time on Murray's farm some months previously, so it was a shock to see the damage when I visited six weeks after the fires. About a kilometre or so from his place, the landscape suddenly turned to sepia. As I drove onto his property, there were piles of twisted, burnt branches and scorched farm equipment everywhere. Ollie was very self-effacing about the events of the day, while Murray was sanguine, despite the enormous amount of damage, the financial loss, the weeks without power and the endless administration dealing with insurance and other service providers. As we walked slowly around the main house, he showed me how close the fire had come, pointing to a patch of tiles lost when the roof caught fire. He praised the firefighters from York and Quairading, who had managed to save his house, even while everything around burnt to the ground.

Amidst all the black and grey, on one native tree that looked completely burnt, Murray showed me a tiny emerging finger of green. He spoke of the optimism such signs of regeneration made him feel. Despite the endless list of jobs to be done, he had carved out time to carefully water his surviving pot plants around the house, that little bit of green helping a lot. He told me about the sunset the night before that was intensely orange. Even though all the trees were burnt, there was still beauty to be found.

Murray has had much support from his friends and family, but what really touched him was the support offered from people he hardly knew. Another silver lining was the blooming, for the first time in about 70 years, of a very special type of flannel flower *(Actinotus superbus)*, which only emerges in the wake of fire. Given no significant bushfires had passed through the region in over half a century, this felt like a minor miracle to Murray, who took great comfort from this evidence of the resilience and regeneration of nature. He invited the Kings Park and Botanic Garden horticulture team to his property to study the flowers and collect their seeds.

Agroecology and Regenerative Farming

In the face of climate change, and recognition of the damage wrought by decades of reliance on fossil fuels, synthetic fertilisers, pesticides and herbicides, a growing number of farmers are implementing regenerative and/or agroecological farming principles. This involves learning from the kinds of relationships and interactions that occur between plants, animals, humans and the environment in nature, and then applying these lessons to agroecosystems. Frequent stock rotations not only allow pastures and soil to benefit from animal inputs before resting and regenerating, but enable close bonds with animals through frequent contact.

There is a lot of simplistic messaging about the negative impact of livestock production. When poorly managed, livestock can seriously damage ecosystems; however, when raised with care, livestock provide valuable ecosystem services. They facilitate the cycling of nutrients, energy and organic materials. Grazing grassland ecosystems reduces the fuel loads for bushfires, while animal manure is a rich fertiliser. Unlike some contemporary commercial breeds that require supplementary feed and, in the case of pigs and chickens, have been bred to suit indoor conditions, heritage breeds have evolved to live in pastured systems, where they can provide these ecosystem services. Moreover, livestock raising is the only viable form of food production in much of Australia, where less than 8% of the landmass is suited to conventional arable agriculture. With histories of livestock damage motivating change, agroecological and regenerative approaches to farming are about building healthy ecosystems that are not just sustainable, but aim to rehabilitate farmlands and support biodiversity within functional ecosystems.

For Greg, the Red Poll breeder we met in the Introduction, his shift to regenerative methods was personal. The doctors attributed his father's leukaemia to exposure to agricultural chemicals. He described how farmers in that era would sit in open tractors spraying highly toxic herbicides on their fields, returning home 'drenched' in chemical sodden clothing. 'It was a pretty hard thing to watch my father die like that', Greg told me, his voice wavering with emotion, 'so that's why we went into regenerative ag, and we just haven't looked back.' Even though Greg does things differently when it comes to chemicals and other inputs, he talks a lot about his dad and his skilled breeding and farming methods.

> He was a very efficient farmer, and that's why he had a lot of success. We probably didn't realise how good he was until he was gone, because there was a lot of people that came, you know, for years and years, to acknowledge what he achieved.

While he was a respected cattle breeder, for Greg's dad, it was by no means just about genetics. Caring for his animals properly was just as important. Talking about some of the challenges and remedies to cattle losing condition, Greg recounted a pivotal conversation he had with his dad on his deathbed:

> What actually took me on this whole thing is I was sitting with my father, you know, in the last week. And he was still conscious, and I said to him, 'what's the best bull you've ever used?' And he just looked up and said 'nutrition'. And he said 80% of breeding great cattle is nutrition and 20% is

> breeding. And I suppose that's what I took from him, and that's what I live by, and I've seen it work.

Good nutrition for grass-fed cattle requires healthy pasture grown in healthy soils. For farmers like Greg, then, caring for the land is key. And this goes beyond just thinking about pasture and cattle health, to looking after the wildlife and native fauna on the farm.

Regenerating Agroecosystems

Unsurprisingly, given their concern for agrobiodiversity, there is awareness and care among heritage breed farmers about biodiversity and conservation more broadly. For the most part livestock farmers pursue their vocation because they love animals and being on the land. Every one of them will tell you that there's a thousand easier ways to make a living. So many nights I've spent on farms across the country have involved sitting outside by the fire, listening to the symphony of insects and animals, under starlit skies, across all seasons. While some are lovely, many more farmhouses are quite dilapidated and used just for eating, sleeping and the essentials. The focus of small-scale farmers is invariably on putting the bulk of their money, time and energy into their land and animals.

Most heritage breed farmers have considerable knowledge about native fauna and flora. Several are involved in formal programs like One Tree Planted and Landcare, or in their own bush regeneration projects. Lindsay on Bunjalung Country in Northern NSW has spent the past 15 years working with his brother towards their goal of restoring 15% of the farm to biodiverse forest. Every year since 2011 they have planted between 500 and 2000 trees. They spend countless hours managing invasive weeds, and have their British White cattle graze the property according to carefully managed rotations. Like other farmers I spoke with, Lindsay believes that while cattle can cause terrible damage to native trees, waterways and land if not well managed, they can also have positive impacts through their grazing. In his words:

> Grassland's not a bad thing and, if you graze it, grazing animals create their own ideal environment. I know that grasslands are not natural here, but we created them by clearing this forest 120 years ago. And I think we have a moral responsibility to use [grasslands] for food production. I can show you when we drive around what happens when you remove cattle. It suddenly goes to shit. Impenetrable, unusable land is created. Camphor laurel, lantana, privet, senna, other major woody weeds that you get...It's basically lost really.

Lindsay is frustrated by the common conceptualisation that 'nature' and farmlands are separate and fundamentally different. Like many regenerative farmers, he sees agro-ecosystems as incorporating native and introduced species in areas of land affected by both natural and human forces. Many native species thrive on farmlands. Willy wagtails are one of the most visible native birds to flourish in proximity to cattle, where they pluck hair for their nests, and flit between bovine legs to catch insects disturbed by their movements. Kangaroos and wallabies benefit from improved pasture, while parrots, ravens and numerous other species raid chook feed, hang around haysheds for feed bounty and shelter, and generally thrive in farmlands.

Livestock, when well managed, can also benefit local ecosystems. In Europe, the role of heritage breeds in managing landscapes is recognised and enacted through the increasingly widespread practice of conservation grazing in wildlife areas. Closer to home on Dja Dja Wurrung Country in central Victoria, Tammi, who started out with Large Black pigs, and who we'll learn more about in Chapter 3, went on to also farm Dairy Shorthorn cattle for land management purposes. She told me: 'We got onto the property, and it was really well suited to cattle. And the slope of the volcano is not suited to pigs. So, we just were like "we need something there to eat down the feed", and that's how the cattle started.' Given Australia's propensity to bushfires, managing grass loads is particularly important.

Another Queensland heritage breed chicken and pig farmer living on Jinibara Country described how his farm exploded with life when he implemented planned grazing, giving areas long rests and frequently moving his animals. In recent years, he has recorded over 100 bird species. (Another farming family told me they had logged over 140 different bird species on their property.) Growing up on the family farm, conservation was all about separation: fencing off a creek here, planting a few trees in a corner there. That's not how he does it now. He sees his pigs and chickens as contributing to the land's health through their grazing and scratching behaviours, and the fertilisation they provide with their manure. Like Lindsay, he sees abandoned properties as in the worst shape with the least biodiversity, as weeds like lantana and wild tobacco take over and dominate. Biodiversity, he believes, can't be promoted through unmanaged landscapes: people should not be thought of as outside of nature, but as integral to it. Pre-colonisation, he points out, First Nations communities actively managed the land for at least 65,000 years to its benefit.

When it comes to land management, heritage breed farmers believe breed matters. Another Dairy Shorthorn farmer in Dja Dja Wurrung Country told me that his cattle are bloat resistant and can handle clover pastures. He said he had far more biodiverse pastures as a result of the resilience of this breed, and his particular bloodlines, which had adapted to local conditions over generations of his family living on the property. Such hardiness is beneficial for conservation grazing and fire load management, as these cattle will graze steeper and higher slopes, and are less selective in the types of pasture they will graze. Carolyn, a Highland cattle farmer in Queensland, enthused about how her cattle 'are fantastic self-grazers, eat anything, but very much want variety. They eat weeds, allowing native vegetation to regenerate.'

Nurturing the land is, of course, not only about giving back to the earth, but also protecting human interests into the future. Carolyn is very concerned about climate change. When I asked her why she first got into Highlands, she answered:

> We looked around for a breed that had longevity, that had gone through numerous types of climates, that were proven to be hardy...You want to have genetics that have stood the test of time, because you've got to start again somewhere if things go really badly.

Conserving biodiversity at an ecosystem level is clearly important to these farmers, but they are also concerned with the loss of diversity in cattle, to which we'll now turn.

A Brief History of Cattle

Archaeological evidence indicates that taurine cattle were domesticated just over 10,000 years ago in the Fertile Crescent in the Middle East. Cattle exhibit the most pronounced breed differentiation of all livestock species. While Asia is home to bovines including the yak (*Bos grunniens*), the gayal (*Bos frontalis*) and banteng (*Bos javanicus*), most of the diversity of domestic cattle has been derived from two cross-fertile species, *Bos taurus* and *Bos indicus,* who share a common ancestor in the now extinct wild aurochs (*Bos primigenius*). From the eighteenth century, the diverse bovine subpopulations were formalised as breeds with the uptake of more systemic selective breeding practices, and the eventual development of herd books and breed societies.

In recent decades, in pursuit of ever greater outputs, cattle have been divided into dairy or beef breeds and selectively bred for milk volume or rapid growth and muscling respectively. Production increases have been substantial yet have come at a cost to animal welfare and breed and bloodline diversity. These challenges are particularly evident in Holsteins, whose phenomenal milk yields have seen their popularity soar. Artificial insemination technology has enabled efficient global dissemination of desired bloodlines.

In the US, over 90% of dairy cattle are Holsteins. In their research into male Holstein lines, researchers at Pennsylvania State University found that, worldwide, the lineage of almost all artificially inseminated Holstein bulls could be traced to one of two bulls born in the 1880s. Their lineages extend, in turn, to two artificially inseminated bulls born in 1960, from whom 99.84% of North American Holstein bulls are today descended. In terms of genetic diversity, these nine million cows are estimated to be equivalent to a herd of fewer than a hundred animals.

Among Holsteins, selection for milk volume has compromised other traits. This has resulted in metabolic and structural problems, increased production disease prevalence, and reduced fertility and longevity in the breed. Janet, a passionate advocate for heritage breeds who has worked on several dairy farms, pulled no punches in describing the harms she sees for dairy cattle in large-scale intensive systems:

> Intensive farming always brings disease in its wake. Anything remotely contagious spreads like wildfire. Not only is mastitis, always associated with overcrowding, rising alarmingly, but other associated lethal illnesses...
>
> There is no place for horned cattle in these squashed intensive dairies with no room to move. No room for other breeds. There is a huge need for water. A huge need for antibiotics. A huge need to get rid of trees for giant irrigators, a huge need for cows to give more and more, an absurd amount per day, at the expense of their health, calving ability, fertility, longevity...The animals have very short lives, compared to family farms.

Along with animal welfare concerns, Holsteins' high milk outputs require high feed inputs, increasing ecological impacts. As Greg put it, 'they're like a racehorse, they need to be fed really well. And they get big performance out of them, but at what cost, you know? You're putting all this tucker in!'

Holsteins also struggle in the heat, with reductions in milk yield when they experience the heat stress that will only increase with global warming. Other breeds, both beef and dairy, are better adapted to warmer temperatures, yet now have vastly diminished numbers owing to the dominance of Holsteins. The Tarentaise, as one example, is less susceptible to heat stress, with milk yield diminishing not nearly to the extent of Holsteins. Incidentally, this breed is also notable for its rich gastronomic history associated with the coveted Beaufort cheese made from raw milk deriving solely from the Tarentaise breed.

The livestock industry's favouring of a diminishing number of the high-yielding commercial breeds has resulted in the extinction of at least 184 cattle breeds globally. This is at least 17% of known cattle breeds. In Australia, twelve cattle breeds are now extinct, with another forty-three listed as under varying levels of threat by the RBTA. Over 80% of the dairy herd is now Holsteins, and approximately 50% of the national beef herd is now Angus (mostly Black Angus, with some Red). This latter figure is striking given the dominance of the more heat-tolerant *Bos indicus* breeds, mostly Brahmans, and tropical composites, like Santa Gertrudis and Droughtmaster in Australia's north.

The Angus' popularity is now self-perpetuating. This is clear in the responses of the Angus farmers I asked about why they opted for the breed. One told me he'd 'not been bothered by the make or model of the mooie, but just stuck to black for ease of access to sales, and appetite for it always being sought after'. Another told me: 'Always had Angus cattle. Reasons for that are because they've fantastic meat, they grow well, they're easy calving. And probably most important of the lot is they're popular. They've been promoted. People know the brand. Well marketed.'

However, as with Holsteins, the Angus has had its share of woes. Intensive selection for profitable traits, and the global dissemination of lucrative bloodlines, has resulted in the spread of genetically determined conditions, including arthrogryposis multiplex and neuropathic hydrocephaly. Fertility has also been an issue with the breed. Joy and her husband have been breeding Angus on Gunditjmara Country in Western Victoria since 1972. After several decades of line breeding, they decided to bring in some new bulls with modern American genetics. The new calves were larger and initially gained weight fast, albeit with much higher feed inputs. However, fertility issues soon arose. 'Because they'd busted themselves producing this amazing calf', Joy told me, 'they wouldn't get back in calf...We discovered then that they only lasted, you were lucky if you got them to their seventh year.' Cows from her old bloodlines would produce calves up to 15 years, and as fertility and longevity are key breeding objectives for Joy, she soon became disillusioned with the modern Angus. 'They're only interested in breeding cattle that have the biggest weight gain. Instead of looking at profit per hectare, they look at weight gain per calf. And they don't look behind what that means', she told me.

In 2012, seeking the hardiness and fertility of the traditional Angus, Joy imported embryos from the native Aberdeen Angus from Scotland. Her cattle are smaller, with less muscling, but they're hardier and graze efficiently on pasture. They calve easily, are excellent mothers and produce premium beef. Joy has spent the past decade working hard to promote the value of the old genetics to help conserve these traits. Her love for her cattle shone through as she told me stories about the habits and quirks of individuals while we walked among her herd in the winter of 2022.

Success Stories: Highland Cattle

While the loss of breed diversity is concerning in cattle, there are also some heartening stories of heritage breed revival. A handful of passionate breeders can really turn a breed's fate around. In Australia, Highland cattle stand out as a shining success story. When the Australian Highland Cattle Society was formed in 1988, their goal was to save the breed locally. Horned cattle, including the Highlands, had fallen out of favour because they can be more difficult to transport and move through yards, and because horns are seen as a danger to people. Proponents of horns in cattle dispute this, however, praising horns for their role in temperature regulation, predator defence, grazing confidence, social relations, grooming and more.

Complex factors determine the success or otherwise of a breed. Given the acceleration of breed extinctions, it is clear that being unique or rare is not enough. Breeds must have special features that make them desirable. For Highlands, like many heritage breeds, they are hardy and produce top-quality beef. Yet, the Highlands have found their popularity largely through their unique appearance. With their shaggy long coats, and dramatic horns, they are beautiful-looking cattle, with particularly endearing fluffy calves. For this reason, they have become enormously popular in the Instagram age among hobby farmers who keep them as 'paddock ornaments'—with benefits. Much is also made of their Scottish heritage, which is important to some breeders, who have made pilgrimages to the Scottish Highlands to learn more *in situ*. Their value has skyrocketed in recent years, such is the demand.

I made my own pilgrimage to the Barossa Valley in South Australia to learn about some of the Highland cattle

Highland cow and calf courtesy of Kym, Smyk Images

conservation work breeders are busy with there. Scott and his mum have a Highland stud and are incredibly passionate about their animals. When I arrived, we spent an hour or so in the paddock, while they told story after delightful story about the quirks of each of their cows. Fights between cattle negotiating their position were the cause of much drama, they explained, but Chelsea, the matriarch, always eventually stepped in to calm things down. The next day, I met up with Scott, his husband, mum, dad and grandmother at a breed promotion event they'd organised: a Highland high tea.

There was a great turnout with about fifty people present. Scott gave an introduction, talking about the breed and introducing each of the cows he'd brought by name, age and genealogy. He welcomed people to pat and feed the cattle celery they had brought for this purpose. He said that, like people, each cow had a different personality, and they'd tell you if they'd had enough. He praised them as a quiet breed, with good temperament. I spoke to the mum of a child on the autism spectrum, who was in the pen with Scott, Evie (a cow) and Connie (a heifer). The little boy was in deep concentration, lovingly brushing Evie. His mum got a little bit emotional as she told me that this opportunity meant the world to him, and that he would talk about it for years to come.

Farm animals are increasingly being used in this kind of way for therapeutic purposes. Katy, a Victorian heritage pig breeder, told me about how she benefits from 'purr therapy' with her cat. She said horses help with anxiety, too, as their heart beats can help regulate our own. She has taken lambs and piglets to schools, who she said particularly delight children with ADHD: 'One little chap nursed a piglet for an hour. The teacher said he normally wouldn't sit still for more than 15 minutes, so the time with the pig gave him a rest from himself.'

Back in the Barossa, we moved into the hall after an hour or so, and Highland beef sausage rolls were served alongside cakes, scones, tea and coffee. The Adelaide-based couple I was chatting to had never been up close to a Highland cow before, and just loved the look of them, so had come along. They were quite surprised to find that Highlands were also on the menu, but they were eager to taste the beef.

The following year, Scott invited me to give a talk on the value of breed diversity at the Australian Highland Cattle Society Show in Mount Gambier. The show served as another promotional event for the breed, where along with showing cattle, they had stalls, information, activities and entertainment, including Highland dancers. I participated in a workshop led by Scott on showing cattle. I was given his beautiful black heifer, Connie, to work with, who later that night over drinks Scott's husband laughingly described as a 'real bitch', to which everyone concurred to my mild alarm.

Fortunately, she went easy on me, bellowing a little, but walking nicely around the ring with me for the most part. I was a little edgy being (ostensibly) in control of this 400-odd kilos of bovine, which wasn't helped by Scott repeatedly emphasising the need for calm, 'as cattle will pick up on your nerves, and get nervy themselves'. At one point Connie trod firmly on my foot, but that was merely a bone-crushing moment of mutual clumsiness. The only real discord we had was when I tried to give her a back scratch. She immediately made it abundantly clear I was crossing a line, despite Greg having showed me the week previously on his farm that this was a core way he endeared himself to his cows. Alas, my attempt at cow-whispering failed, and

Highland cow and calf courtesy of Erica Smith

Connie headbutted me and made it very clear I should keep my hands to myself.

Every night of the Highland show, over dinner and drinks, the breeders spent hour upon hour gossiping about individual animals. I was bemused and delighted by the intimate knowledge the South Australian contingent had of each other's animals, and the level of intrigue and fascination on display as they unpacked the latest breeding dramas, fence abuses, bull face-offs and other misadventures. Another passionate Highland breeder, Erica's pride was palpable as her bull, who had taken out several ribbons that show, was praised effusively. While conservatism prevails in some agricultural communities, the Highland mob are particularly inclusive, with a gay president, an older woman as the show's judge, and predominantly female stewards, handlers and winners. This is something they take pride in, and a trophy was awarded on the night of the final dinner that celebrated inclusivity in their community.

Loving Cattle

It's easy to see how the Highland mob become so enthused about their animals. Cattle are playful and lively beings. They are fascinating to observe, and I have spent many a happy hour, initially during childhood adventures at my uncle's dairy farm, staring into paddocks ruminating about ruminants. I love watching mothers lovingly lick their calves, leaving scruffy wet patches all over their fur. Calves energetically cavort together, running around, playing and mounting each other mischievously. Sean, an Angus farmer in Menang Country, has a swing hanging from a large Yate *(Eucalytus cornute)* tree that he had hung for his kids. The cattle love it and will frequently push it, chew on it and play with it. His huge bull looks particularly comical playing with the swing. Cattle are endlessly inquisitive and will play with anything they find, particularly unusual things, like stray anthropologists. The first time I visited, Sean had me lie down beside him still and quiet for about half an hour in the middle of a muddy paddock while his mob of at least thirty cows surrounded us in a jostling scrum, sniffing at us curiously, while giving the odd lick.

Not far down the road lives Alister, who at the time of my visit was ninety-four. On a rainy winter's morning in mid-July 2021, he took me out in his ancient ute to meet his Sussex cattle. We were chatting as we drove about cow intelligence. He recounted how just a couple of days prior, his mob had worked out how to open a gate to get into the hay shed. He put them back in their paddock, but sure enough they opened the same gate in the same way the following day to get back to the hay. Driving through a far paddock, the herd spooked and bolted into the distance. We drove back around towards where the cattle had run, but they weren't where he hoped they'd gone. I said, 'that's alright', to which he responded, 'No it's not! I like showing off my cattle!' We caught up to them eventually, and Alister talked me through the breed's traits, and the individual quirks and histories of the cows around us, with great enthusiasm.

Of his Red Polls, who have a similar colouring to the Sussex, Greg is similarly passionate. He talked at length about their temperament, which he is very proud of:

> They've just got a kind eye, you can pick them with your eye, a kindness in there. And my mum used to talk about kindness in the eye. You can see it in horses, you can see it

> in cattle. Yeah, and their ears, and how they hold their head and that. And we sell probably fifteen to twenty bulls a year, and we can tell just by their nature…And people just come back and say they become in love with their animal, you know, and look forward to seeing it.

Farmers like Greg frequently articulate that they see their animals as family. In this vein, in giving advice to a new farmer on habituating a cow and her calf for milking, a more experienced Dexter breeder offered the following to a new breeder:

> Before birth, teach Mum to be fed in a yard, so she gets used to the routine. Handle her often. Never get between the cow and calf. Don't forget to spit in the calf's mouth to imprint your scent. That helps, and everyone is family then.

Animals form a core part of the social world of farmers. They are the beings they spend much of their time with. Farming is a vocation that you live and breathe and defines very much who you are, often over many generations. All of Greg's siblings are still involved in farming, for example. Of going to show their cattle at agricultural shows of a weekend, he said, 'it's like playing football or netball on the weekend, you know, it's a bit of fun'. His two girls are immersed in the world of cattle, and for Christmas during the first year we were in touch, Greg told me they were super excited that Santa had given them tickets to the Stud Beef Victoria Handlers Camp at the Geelong Showgrounds.

More specifically, farmers become strongly identified with the species and the breeds they raise. Highland cattle are rare in the Darling Downs where Carolyn lives, and she and her husband are known about town as 'the hairy cow people'. Greg will be a Red Poll man for life even though this comes at a cost. He earns considerably less at the saleyards per animal than he would if he succumbed to the Angus trend, and he constantly cops a ribbing from his community about this:

> You see this massive drop [in price per kilo] and you feel like a leper. It's a terrible sinking feeling, and you might know some people at the saleyard, and they go, 'oh Greg, you should have just put Red Angus on them, and you would have made top dollar'.

But he never would. Greg perseveres with promoting Red Polls, who he passionately believes have the best maternal characteristics, produce excellent beef, and are cheaper to raise, with lower inputs and fewer vet bills given they are so hardy. He sees the Angus game as a false economy.

In thinking about the respect and love farmers have for their cattle, I became very interested in cattle–human communication. Following a discussion about one of his 18-year-old cows, Tom went on to comment:

> There's something about a slow, old cow that I love. I just love them. I love them. They're a thing of beauty to me. A big, old, slow cow that's just doing her best, pottering along and will come up and look at you as if, 'well, when are you going to give me some food? And why are you wasting your time with that even?'

Along with such intuited communication, farmers talk to their cattle and their cattle vocalise to them. When cattle feel there is insufficient feed in a paddock, for example, they will moo loudly and persistently at the gate until their farmers move them to fresh pastures. In turn, when they need their cattle to move, farmers call their cattle with a summons familiar to anyone who has spent time in the country: the tried-and-true, Aussie-accent-inflected call of 'c'mooooooorn!' Interestingly, a study in a British dairy found that cows who had names produced a significantly higher amount of milk each milking than those who didn't. Talking to cows and sharing camaraderie of sorts with them appears to affect contentment and even productivity. That said, farmers are at pains to emphasise that these relationships are not the same as those they have with their horses, dogs or cats. As prey animals, cattle have a natural wariness of humans, which results in different kinds of connections, communication and behaviours.

Danger from Cattle

Good temperament and positive relations with cattle are important given their size and the danger they potentially pose. In the years between 2001 and 2020, cattle killed thirty-one people in Australia. Indeed, farming more broadly has a terrible safety record relative to most vocations, with high rates of death and serious injury. Protective instincts can lead cows with new calves to be aggressive. Doreen, who is married to Alister, described:

> I can remember Alister getting chucked up in the air like a ragdoll with a cow, and our son got the same. But both times it was a cow that was freshly calved, and you did something wrong. I had a cow and a calf, and what

> I should've done was went and got the calf, and put the calf up here, where the cow could see it. But when she got up here, she found the calf wasn't there, and then turned round and took it out on me. But, again, that's something you don't get taught, you learn, sometimes the hard way!

While most cows habituated to humans are easy to handle and pose little danger, as Doreen described, even the most placid cows can become dangerous with a new calf at foot. Bulls can also pose a risk and a common mantra among cattle people is 'never trust a bull'.

Among habituated cattle generally, counter to what one might assume, the most dangerous bulls tend to be those hand-reared by humans in situations where they are separated from other cattle (this holds true for rams, too). It is theorised that, if not raised and socialised with other cattle, they imprint on humans, triggering species confusion. This means that when they reach sexual maturity, they see other humans as sexual competition, so engage in their natural behaviours of tussling and fighting for dominance.

This is well known to experienced farmers, and they worry about those new to farming who might feel treating livestock as pets would have positive behavioural outcomes. It is a fine balance where, ideally, livestock are habituated to being handled by calm, kind, communicative humans who also retain a level of authority. Regarding the latter, given the hierarchical nature of bovine sociality, asserting dominance is seen as key to safety. Greg described his father proudly in this respect:

> My father was a real bull wrangler. He would go down into the Jersey bull paddock with a pitchfork. Yeah, there could

> be a dozen or more bulls in the paddock, and they'd be out pawing at the ground and breathing fire. And we'd sit on the rails, you know. If the bulls ever came at him, he used the pitchfork as, like he sort of put it in the side of their flank, and he sort of rode them around in a circle. And then the bulls would realise that he was boss, and he would just walk out whichever bull he wanted out of the paddock. There was only a little gate about 3-foot wide, and he'd walk the bull out of the paddock, and the bull would start heading towards the gate, because he knew dad was the boss.

Of course, in the grand scheme of things, cattle are in infinitely more danger from people than we are from them. While there is considerable annual variation, around 6 million cattle are slaughtered each year in Australia alone. I was present for a home kill of Sean's two-and-a-half-year-old Angus heifer. She had failed to ever get in calf, so was chosen for slaughter. Sean was full of nervous energy on that cool overcast morning. When the time came, he strode purposefully into the paddock where she was lying down ruminating. The crack of the rifle echoed out across the paddocks, and his neighbour, a trained butcher, rushed to slit her throat and commence bleeding her out.

While killing is a normal part of seasonal life on the farm, Sean does a home kill only once or twice a year, so while there was joking and camaraderie in the process, there was an intensity to the mood. There is no turning away from killing here; it is visceral, it is bloody, and it is hard work, both physically and emotionally.

I drove back to his farm ten days later after the heifer had been hung to ensure tender, flavourful meat. We spent the best part of two days butchering the carcass and, along with all the usual cuts, made mince and sausages. We cleaned out his deep freezers and labelled, stacked and froze the huge amount of meat. Sean was tired, withdrawn and a little grumpy at the end of the first long day. We talked a little about it, and I made an inane comment about his feelings. He countered, 'It's fine. If I stop feeling something, then that's the point where I no longer deserve to be a farmer.' Farmers see this kind of emotional engagement as underpinning an ethic of care that ensures a humane approach to raising and killing animals.

The small-scale farmers I have come to know take a lot of pride in raising their animals with care. In making sense of both loving and killing animals, they frequently comment that their animals have a great life and just one bad day. Farmers are aware, of course, that this one bad day shortens their animal's natural lifespan considerably. In grappling with the emotional complexity of killing animals they love, farmers justify their practice through looking to cycles of life and death in nature, and to predation, particularly. They also talk frequently of what they see as the grim conditions of intensive livestock production, which they contrast with their approach of high levels of care and intimate killing that they see as more ethical.

First Nations Connections

We will end this chapter with the wonderful story of a Ngambri man and a heritage breed cattle farmer who are working together to progress social and ecological justice. In the Introduction, I talked about the central role of livestock farming in the colonial dispossession of First Nations people. Notwithstanding this brutal and violent history,

Aboriginal people's resilience has been extraordinary and many have become involved with livestock raising. In the country's north, particularly, many Aboriginal community members adjusted to life on cattle properties where they became expert cattle handlers.

Today, First Peoples continue working with cattle in various capacities. In the course of this project, I did not meet any heritage breed farmers from First Nations' backgrounds. However, in the past few years, the RBTA has been in contact with an Aboriginal corporation wanting information on banteng cattle, while elders from the Coburg Peninsula have been involved with efforts to save the Timor pony.

I have, however, had many conversations with non-Indigenous heritage breed farmers about farming on unceded lands, both today and in the past, and their connections, or at times lack thereof, with local Aboriginal people. When I asked about this, one farmer told me that the local people 'didn't like' the place where he lived, implying there was some kind of taboo around his land. This farmer was a first-generation migrant, so his family didn't historically have direct involvement in colonial dispossession; however, it was clear he found some level of reassurance in this idea, which appeased any discomfort he might have otherwise had. The claim itself may or may not have been accurate. The broader region certainly has a long history of First Nations connections.

Other heritage breed farmers are as engaged with social justice as they are with ecological reparations. This cohort expresses great frustration with the slow progress of reconciliation nationally. On the unceded lands of the Ngambri and Ngunnawal people on the Southern Tablelands of NSW is a family running Belted Galloway cattle, and producing meat and eggs from Silver Grey Dorking chickens, along with fruit and vegetables. The farm's current custodian, Murray (not to be confused with Murray from WA in the opening passages), grew up witnessing the infliction of some of the wounds of settler colonialism. His father was a police officer, and he described the culture of entrenched racism he witnessed within community policing and the justice system. He credits his wife for helping open his eyes to what he sees as the 'unfinished business of this nation', of righting the wrongs of settler colonialism.

Reading Charles Massy's *Call of the Reed Warbler* catalysed Murray's journey into regenerative agriculture. Massy's emphasis on land stewardship impressed upon him the idea

Belted Galloway cow with one-week-old calf courtesy of Scott Carter

that he didn't own the place, but was a temporary custodian. This led Murray to consider that, if it's taken 200 years to 'stuff the place up', it's arrogant to think he will have all the answers on how to fix it. He really wanted to explore Aboriginal land stewardship, which led to a much deeper and more profound journey.

After arriving at the farm in 2018, the process of connecting with the traditional custodians was not straightforward. The Ngambri and Ngunnawal traditional owners were alienated from their lands during the colonial period, and the histories of dislocation, disadvantage and intergenerational trauma continue to have resonance today. After considerable engagement and a number of false starts, Murray eventually began to build a connection with Ngambri traditional custodian Paul, who, in time, came out to the farm. For Murray it was a profound and moving experience to be with him on his ancestral lands: a place of great significance for Paul's family and community. There were tears, Murray recounted, and it was an emotional time. Paul impressed upon him that, as the current custodian of the land, Murray needed to care well for it for the whole community and future generations.

The relationship of respect and trust developed into an ongoing partnership conceived through the lens of *yindyamarra*, a Wiradjuri and Walgalu term that means 'mutual respect', and the actions and responsibilities that go with that respect. Murray told me that this entails a way of being that recognises our duties to each other and the environment. He said that in its broader meaning, *yindyamarra* refers to a way of life and philosophy based on responsibility, respect, wisdom and an understanding of the interconnectedness of the natural and spiritual world.

A key part of their partnership involved the returning of a portion of Murray's family farm (approximately 12 of their 89 hectares) to Paul and his family. The land is for them to care for and use however they want, under whatever time frame they deem appropriate, with any derived revenue remaining with Ngambri mob. Plans include a tourism venture focused on sharing traditional knowledge and practices, such as tree scarring and on-country ceremony. Paul is also considering an agroforestry experiment in the creation of original timbers used for making traditional implements. A commercial crop of *myrnong* (yam daisies) is also anticipated.

Murray described how the partnership has gifted him with a very different way of relating to the land. As someone not from an intergenerational farming family, he said he lacked a deep understanding of seasonality. He feels he has often been on the back foot, responding reactively rather than proactively to seasonal change. Spending time with Paul allowed him to learn a new sympathy for and understanding of the land. He looks now through a 'do no harm' lens. He talked of picking up 'the heartbeat of the place' and learning to think deeply about what the land needs, largely through spending a great deal of time just observing. 'Those things, if you make time, are pretty powerful', he commented. 'Without Paul, we were rushing around like madmen doing things, and we were missing a lot.'

Ultimately, however, land-sharing has proved to be only a stepping stone towards other partnerships for Murray and Paul. While it serves to build connections around co-stewarding, it is limited by the fact that there can only be one name on the title deed. In this respect, despite mutual best intentions, they see it as continuing colonial structures as it fails to redress the ownership of stolen lands. What is needed is a new way, which Paul calls '*Marra marra murru*' (creating pathways). Murray believes that:

> central to meaningful progress for First Nations people is their absolute right to economic self-determination. If we are to begin to address the underlying issues caused by displacement, we must accept that without freehold title land ownership, First Nations people will not achieve their economic independence.

Together they have created the Waluwin Foundation, an organisation that Murray told me will show a new/old way for First Nations nation-building, starting with support for the Indigenous agrarian economy. This, he said, is an economy built on *yindyamarra*, an economy that respects Mother Earth and that has been in place for tens of thousands of years.

Waluwin will offer farm loans: *Ngama bangalnarranarra*, Mother Earth loan. This loan aims to allow First Nations people to return to Country as farmers on their own freehold title land. It is farm finance that seeks to deliver First Nations economic development and self-determination, along with improved soil health, biodiversity and animal welfare, and reduced loan anxiety and bankruptcy. Paul and Murray's collaboration has the goal of *Waluwin mayiny* (healthy people) and *Waluwin ngurambang* (healthy Country).

Conclusion

Over the past two centuries, settler-colonial land management and livestock production practices have alienated First Peoples, caused ecological damage and contributed to global warming. Responding to these histories, the agroecological and regenerative approaches pursued by many heritage breed farmers have the potential to feed humanity through enacting respectful and ecologically sustainable practices that foster biodiversity at multiple levels. Heritage breed farmers could make more money by raising the more profitable and readily available commercial breeds, yet their love for and commitment to their animals is the cornerstone of preserving bovine breed diversity. Nurturing the land and animals is also, for some, extending to seeking to progress social reparations. Through working together, First Nations community members and heritage breed farmers offer a hopeful and practical model of moving forward towards a more just and equitable future.

Chapter 2

Sheep

Tom's family has lived in the rugged highlands of Dja Dja Wurrung Country since the 1870s, when his ancestor, Edward, turned the family's fortunes around on the Ballarat goldfields. Hailing from a Cornish tin-mining family, and unable to write anything but his name, Edward's gold haul enabled him to move from the hollow tree that served as his first Australian home to the house he built on 53 hectares he secured from the Crown in the high country. Edward went from mining to feeding miners; growing potatoes and berries, while planting out an orchard that remains on the property to this day. Then 'at some point lost to history', as Tom put it, his family started running traditional Hereford cattle, whose British genetics are distinct from the North American Hereford bloodlines that he says predominate in Australia today.

Tom's mum also derived from a long line of cattle farmers. She passed away some years ago, but her heart had always lain with the Dorset Horn sheep that her family had raised for generations. Tom speaks with pride of his mum's extraordinary understanding of her sheep, with her expertise much admired in the community by other farmers and even the local vet, who would contact her for advice. He described the special relationship his mother had with Tess, one of the Border Collie sheepdogs the family kept. Tom's dad was also inseparable from his dogs, but Tess and his mum worked differently. It was 'as if by telepathy' according to Tom, who says it was incredible to watch. Tess could pin and hold a lambing Dorset Horn ewe—no mean feat given their size and strength—so that his mum could come to the ewe's aid in the case of a difficult birth.

Like their beloved Dorset Horns, Tom describes his Border Collies as family. When he was two, he wandered off into the dense bush on their property:

> The neighbour rang up on our telephone line and said, 'are you looking for Tom? Well, Shep is bringing him up the hill.' And it was wild country. And so the dog had left Dad—which he never did, never left Dad—and the dog was there with this two-year-old kid. The neighbour said 'the dog is bringing him up the hill. Nose either side, shepherding him up the hill, bringing him home.'

As a confessed romantic, Tom loves this story of being herded like another errant sheep. His life and memories are measured in epochs determined by which of the four dogs of this bloodline he had at the time.

Tom's story gives us a taste of the deep entanglements of sheep, sheepdogs and humans in Australia, where each has shaped the other's trajectories and the landscapes in which they live, over generations.

A Brief History of Sheep

Sheep were among the first animals to be domesticated for their meat, milk and hides around 11,000 years ago in Mesopotamia. Wool sheep were developed more recently at around 6000 years ago. Today, there are around 1.25 billion sheep scattered across diverse environments globally and, according to the FAO, over a thousand distinct breeds.

In 1788, enroute to Botany Bay, the First Fleet took onboard local hair sheep, thought to be the Afrikaner breed, in Cape Town, South Africa. These were large-framed sheep, whose tails grew so fat they had trouble naturally mating. They were soon supplanted in the colony by other breeds. Sheep were initially housed in makeshift barns overnight, but settlers soon realised that the weather in NSW was sufficiently mild that sheep could be left out overnight.

That sheep and cattle could flourish with less labour and management enabled the vast expansion of pastoralism in the new colony. In the fenceless expanses, shepherds, usually former convicts, led a meagre existence herding flocks of hundreds throughout the year. Systematic breeding was impossible in the absence of fencing, so few 'improvements' were made through selective breeding in the early years.

The impact of sheep on Australia's colonial development cannot be overstated. For the first few decades of settlement, the hungry colonists valued them more for their mutton than their wool. From the 1820s, however, the annual wool clip became the primary source of the nation's wealth. The booming industry resulted in the inland pasture lands being increasingly coveted, triggering violent conflict between settlers and First Nations communities. Hostilities erupted when First Nations community members defended their land or killed sheep and cattle, or when settlers damaged waterholes and killed native animals. Along with executing brutal massacres, as they moved into the interior, the predominantly male settlers abducted and raped First Nations women. Over time, as the need for labour increased alongside the growing flocks, First Nations men, particularly, were recruited as shepherds, shearers, sheep-washers and reapers, oftentimes poorly paid, if paid at all.

Along with these social harms, as sheep numbers exponentially increased, they compressed and damaged the continent's fragile, ancient soils. Counter to what you might expect given their modest stature, sheep can be particularly hard on land, especially when overstocked. This is because they graze pasture very low. In WA's arid Wheatbelt, Linc, whose family has for decades run cattle and sheep, described how:

> If you have a herd of cattle walk across the sandplain, and a flock of sheep, if you look afterwards, where the flock of sheep have crossed, that land is really opened up to wind erosion and stuff. Whereas where the cattle have walked across, it is not as disturbed, even though cattle are bigger.

Notwithstanding their impact, pastoralists grew sheep numbers and by the mid-twentieth century, Australia was home to the most lucrative wool industry globally. However, by the turn of the century, the industry had imploded. Human-made fibres and cotton proved stiff competition, but commentators suggest it was the politics, power struggles and failure to adapt to a changing society that ultimately tore the industry apart. Today we produce only a third of the amount of wool of the industry's heyday.

Extinction

Globally, at least 14% of sheep breeds are extinct already, with many more under threat. The RBTA lists 14 sheep breeds as critical in Australia (less than 300 breeding females), four more as endangered (less than 500 breeding females) and five as vulnerable (less than 900 breeding females). The sheep type in most jeopardy in Australia is the carpet-wool breeds. Wendy has the only remaining flock of Drysdale sheep in Australia. I visited her during lambing, and it is astounding that such a beautiful breed is in such jeopardy.

The Tukidales and Elliottdales are also in dire straits. There is only one flock of Tukidales in Australia, and in their native New Zealand they have been declared extinct, so the future of this breed is bleak. In Australia, we have lost six breeds completely, including the Cotswold, Oxford Down, Polwarth (horned), South Dorset Down, Teeswater and Wensleydale. Yet, people still remember and commemorate these lost breeds. As one example, in a sheep-related Facebook group, one person posted an image captioned:

> The beautiful Cotswold sheep. We had this stunning breed in Australia for around one hundred years. They were brought out in 1826, by the Van Diemen's Land Company and seemed to find the most popularity in Tasmania, where the last one was seen at the Launceston Show in 1913.

Even while some locally extinct sheep breeds live on elsewhere, because of diseases like scrapie (that we thankfully do not have in Australia), it is prohibited to import live sheep, and it is very complex and prohibitively expensive to import ovine semen and embryos. That said, in recent years, some breeders have jumped through the various

Dorset Down ewe and lamb courtesy of Henry Dunn

bureaucratic and financial hoops in order to establish a flock of the purportedly cutest (and now the most expensive!) of sheep breeds, the Valais Blacknose.

To return to the Elliottdale, this endearing and highly endangered carpet-wool breed has a uniquely Australian history. The breed was developed in the 1970s at the Elliott Research Station in Tasmania for the carpet-wool industry. Each year they grow around 30 centimetres of strong, coarse wool of at least 40 microns, which exhibits the springiness that makes their fleece perfect for carpets. They have good growth capacity, tasty meat, impressive fecundity, and strong resistance to foot rot. In their heyday in the 1980s and 1990s, there were over a hundred breeders and tens of thousands of Elliottdales being shorn twice annually in a promising emergent Australian wool carpet industry.

By the turn of the century, the bust superseded the boom, and today there is only a handful of remaining flocks. One is owned by Sue, who works passionately to promote the breed through social media. Some of her quirky individuals, such as Dottie, now have a minor cult following. Sue was sheep coordinator of the RBTA at the time when the breed's major proponent developed cancer and had to disperse the flock. She told me:

> We tried all sorts of things to get people to take them on. Couldn't do it, because there's no money in them. They have to be shorn twice a year, they're a carpet-wool sheep. Nobody buys the carpet-wool anymore. So, it's basically a conservation flock. Elliottdales were developed here, so if they go, we can't get them back.

Manufacturing of wool carpets ceased in Australia for complex political and economic reasons, but among the challenges was the rise of synthetic carpets. Produced at scale in China, they are considerably cheaper than wool carpets. Synthetics are popular also because of durability and ease of care. With clothing, for example, wool often has to be handwashed, whereas synthetics can just be thrown into the washing machine.

Consequently, apart from Merino wool, which we will turn to in a moment, wool has lost its value, and farmers, at times, can't even give it away: the cost of shearing is often higher than their earnings from the wool. A further challenge is that Australia is struggling with a shortage of shearers. Today there are only around 2000 shearers nationally, whereas in 1990, there were almost 10,000. Breeding values have meant ever increasing sheep sizes, where sheep can exceed 100 kilograms, as farmers pursue ever more meat and wool per animal. Upright shearing platforms aim to mitigate shearer injuries and discomfort, but the technology has changed little over the decades.

I was at Judy's, a Queensland-based advocate of rare breeds of all species, when her shearer came to attend to her thirty-four sheep of various breeds, including English Leicesters, Babydolls and Dorset Horns. It is incredibly physically demanding work, undertaken at an uncomfortable, back-bending angle. The shearer was sweating profusely after the first few sheep, but raced through them all, taking two to three minutes each, without a break. Afterwards, we were chatting, and he mentioned that his daughter had Covid. I asked if all the family had come down with it, and he joked that he always felt achy from his work, so he hadn't bothered testing himself.

Many hang up their shears by their mid-forties due to bodily wear and tear. Shearing, and sheep handling more broadly, can also lead to injuries. One meme on sheep social media had an accompanying image of a collage of nasty bruises on people's legs and was captioned: 'tell me you own sheep, without telling me you own sheep'. Comments included one from a woman who described how when she was shearing, she went to visit a relative in hospital, and a nurse pulled her aside to ask if she felt safe at home.

Because of the low value of heritage breed wool, shearing challenges and the shearer shortage, the shedding breeds that don't need shearing, such as Dorpers and Wiltipolls, have become increasingly popular. Yet, notwithstanding the current situation, some heritage sheep breeders are optimistic that growing concerns about sustainability will bring natural fibres like wool back into favour. There are certainly many creative uses people have been putting wool to in recent times. Wool is being used as a biodegradable form of packaging by some companies in Australia, as mulch in orchards and gardens, and, famously, for Prince Philip's biodegradable coffin.

Merino Sheep

When it comes to relatively consistent market value, Merino wool is the exception, due to its fine, soft nature, which makes it suitable for skin contact. During the period of the Spanish Merino's ascension, its association with the elite in British society served it well, while the traditional British breeds came to be associated with working-class farmers. Arriving in Australia in 1797, the Merino's fine wool was highly coveted by English wool manufacturers and led to the hugely lucrative industry discussed earlier. The rise and rise of the Merino is a fascinating history, which I don't have space to go further into here. The key point for us is that, like Holsteins and Angus in cattle, their popularity has affected the survival of other breeds.

While Merino wool is certainly a beautiful product, it can't fulfil the functions that other coarse wool types have done historically, and some believe the breed is disproportionately valued. Jenny, who lives on Yorta Yorta Country, and has both Merinos and heritage English Leicesters, explained:

> It's the Merino machine, I call it. It's all advertising, you know, the Merino is pitched as—even though my husband runs Merinos—it's pitched as something that to me isn't all it's cracked up to be. You cannot use Merino for a pair of socks unless you put something else in it to stabilise the wool, because it wears away. It's too fine...making blankets, and all those sorts of things, they're all gone, pushed aside... I'd like to see that change, and we go back to looking at the individual traits of all of the different breeds. They were here for a reason.

Merinos are also more susceptible to flystrike than heritage breeds because of their wrinkly skin and dense wool. Consequently, they are often subjected to the painful (and polarising) process of mulesing—cutting off flaps of skin around the tail—in efforts to prevent flystrike. This is a painful condition where remnant excreta attract blowflies that lay eggs, whose larvae then eat the flesh of the sheep around the breech, tail and/or belly area, causing open wounds and infection that can prove fatal.

In wet conditions, if fleece rot occurs, then body strike can follow across the sheep's shoulders and backline. Flystrike is one of many health and welfare issues plaguing Merinos, who of the various breeds tend to be thought of as the least robust. A cartoon of an exaggeratedly large book captioned, 'How to Keep Sheep Alive Volume One' circulated through ovine social media last year. Among the most popular comments was '99% of the book would be on Merinos', attracting many likes and affirming comments. Someone then replied, to much knowing amusement, 'the remaining one percent would be about your best ram'.

As this latter comment indicates, it is not just Merinos who suffer health challenges, and sheep are considered prone to sickness and misadventure. In coping with this perennial challenge, there is ongoing humour among farmers about the species' love of dropping dead. On a popular sheep Facebook page, for example, an image of a sheep staring into the camera was captioned with the challenge to comment on what the sheep was thinking. The vast majority of comments were along the lines of: 'Definitely thinking about the most embarrassing/expensive way to attempt to die'.

A rather banal way that sheep expire is by getting stuck on their backs. A so-called cast sheep often can't get up without help and will die if not discovered in time. Sheep carrying a full fleece or pregnant ewes are most susceptible. Driving around Brenton's English Leicester flock in Tommeginne Country in Tasmania, we came upon a cast ewe, which was the cause of some surprise given his mob were recently shorn. Brenton rolled her over and gently held her head up for a minute or so. He left her in a sitting position, while phoning his son and asking him to check on her again before going home that night. He then got her on her feet, and despite looking pretty floppy and weak initially, she soon wandered off and reintegrated with the flock.

Heritage Breeds

The many and varied breeds of sheep possess different qualities, but their advocates generally praise their fertility, longevity, hardiness, meat and/or fibre qualities, and cultural heritage value. Since 1986, Rob has had Cheviots, a breed who derive from the hills of the same name between Scotland and England. He describes this distinct pointy-eared breed as hardy and able to tolerate the wet and hilly West Gippsland climate on Kurnai Country, where he often receives over 1000 millimetres of rainfall per annum. 'The sheep can literally be standing on soft, muddy ground for months, without the need to touch their feet due to their soundness', he observed. Given the increased rainfall some areas on the east coast have been subject to in recent years, these qualities may become increasingly important.

Shropshire sheep enthusiasts often choose their breed for organic weed control given their ability to graze among established trees without ringbarking and killing them. They are hardy, disease resistant and docile, and produce abundant milk that can support twins and triplets. Ewes will continue to lamb often to ten or more years, usually without complications.

Phil runs Shroppies (as they're affectionately known) in his cherry and lemon orchard in Peramangk Country in South Australia. When they transitioned to organic farming, a core challenge was how to manage the weeds and grass in the orchard. Mowing is tricky with their reticulation system, while organic herbicides are expensive, labour intensive and not completely effective. Phil's Shroppies will eat foliage within their reach, but their small stature means they can only reach the lower branches of the trees, so they don't cause significant damage. And they're lovely to have around. In Phil's words:

> They will come when called (trained with lucerne), are easy to handle, are good mothers, gentle on fences and seem content to stay within the bounds of our orchard (which I believe they view as their home). Their manure is a welcome addition to the nutrition of the trees—our farm has never looked better.

Shroppies have been in Australia for over 150 years. In 1903, there were 140 registered studs, and they dominated the prime lamb market. The past century, their numbers have sharply declined. Through the work of some passionate breeders, however, in recent years, a growing number of orchardists have come to recognise the role they can play in weed and pasture management, and their popularity is growing. In 2019 there were just ten registered studs, but by 2023, this had risen to 25.

Shropshire breeders rave about their meat, which is valued for being tender and juicy. Kyla is a 36-year-old farmer and butcher living on Noongar *boodja* in WA. Her parents had the classic love story of shearer meets roustabout, so raising sheep is 'in the blood'. Kyla had this to say about the breed's eating quality in an article in the RBTA newsletter:

> It may seem morbid but as is often said, the best way to conserve a breed is to eat them. I run a small Shropshire

> stud mob, and I always joke that they are the perfect butcher's sheep. Tiny legs, tiny heads, and a whole lot of meat. As we move away from large, nuclear families, the need for large cuts has reduced as well. Consumer taste has moved away from huge and bone in cuts, and more towards smaller, value-added meat products. This is part of why I choose to run a smaller sized heritage breed.

Perhaps unsurprisingly, I have noted a pattern where those who have come to farming as adults are more likely to struggle to eat their own animals than intergenerational farmers for whom it has long been normalised. Kim and Geoff live on Dja Dja Wurrung Country and breed critically endangered Finn sheep. They love lamb as a meat, but when they left the city and bought the farm, they chose a fleece and milk breed that they wouldn't have to slaughter. Kim explained:

> I guess what it is, is that you see them come into the world, and you watch them, and you fall in love with them, and they've got all their quirks. And then if they get sick, I treat them according to what they need…And that little creature that trusts you, then to send it to slaughter, I just couldn't do it.

Not far from them is Sue, who also came to farming as a second career. Along with the Elliottdales, she raises another four breeds of heritage sheep and is heavily involved in rare breed conservation through volunteer work with various societies. She doesn't eat lamb, but she has just never liked it. Her thoughts:

> I don't have a problem with people eating lamb. And I think that the best way of eating anything is what you grow yourself. You know it's had a good life until that one bad day. And nobody can really hope for more than that. Regardless of whether we're human or not, if you have a great life, and then the one day you die—okay.

Jenny eats her English Leicesters and believes 'the meat taste is a whole lot better than just a plain old Merino. Merinos are pretty bland.' She has also been super impressed with her English Leicesters' relative longevity. Jenny and her husband have been on their farm for 30 years and currently have 3000 Merinos that they run commercially. On her fiftieth birthday, Jenny decided she wanted to 'do something altruistic and get some rare breed sheep', and so she bought some English Leicesters. She was struck by their longevity:

> A lot of the more modern sheep breeds, they tend to be really vigorous for the first couple of years, and then they peter out, so it's a bit like the throwaway society. You know, every couple of years, you have to keep replacing all your rams. But I'm finding with the Leicesters...that longevity is there.

For Tom, what he finds so special about his beloved Dorset Horn sheep is the way they respond to people. In his words:

> The difference is they are a flock that requires human interaction. They have evolved exceptionally closely with

> humans...Because of the climate, and them being kept indoors and handled so closely, they have a much closer relationship than the Spanish breeds...It is uncanny the way the lambs interact with you. We were bottle feeding... You would talk, and the Dorset Horn would look at your eyes, would look at you, whereas the other sheep would just be this scattered thing. And you walk over, and they've all been bottle-fed, you walk over, and the other breed scatters, and the Dorset Horn comes up to you...And I've worked with Polwarths and Merinos, and the behaviour is just so drastically different.

He went on to describe how you can walk up to a recently lambed ewe in the middle of a paddock and pick up her lamb. She won't attack, she'll just talk reassuringly to her lamb. He says that if you, 'pick up the lamb and walk with it, she will follow you. They will turn you vegan! They are just amazing.'

Dorset Horns have other interesting characteristics. They have an aseasonal, or extended, breeding season. Some even have two lambings a year, with multiple lambs not uncommon. This can provide a niche in the market, as lamb can be sold when other breeds are not producing. They cope very well in the heat, and their horns have historically been used to craft tools, such as cutlery, and walking stick and hairbrush handles. Horn is incredibly durable and lasts centuries. However, like so much else besides, plastics have replaced this more sustainable material, and horns mostly find a market today in niche manufacturing of dagger handles and the like for medieval re-enactment circles. It should also be noted that the Dorset Horn has the rare accolade of big screen success, having featured in *Rams*, the Australian remake of an Icelandic film.

Finally, on heritage breeds broadly, while sheep are saddled with the cultural cliche of being daft, heritage sheep breeders take umbrage at this. They argue that the older the breed, and hence the closer to its wily wild relatives, the more intelligent it is. They describe sheep as readily capable of learning and passing on information within the flock. 'Hefting' is a term used in Britain to describe a flock of sheep's knowledge of and attachment to their territory. Hefted sheep know how to safely navigate tricky mountain terrain, where to seek shelter in poor weather, what to graze, where to locate water and the intricacies of farm life and infrastructure, such as the location of gates, routines and patterns of movement. When authorities implemented mass culls of sheep during the devastating foot-and-mouth disease outbreak of 2001 in England, this knowledge was lost. While sheep were replaced, the new flocks' lack of knowledge made farmers' lives far more difficult.

Loving Sheep

Like the other livestock species discussed in this book, sheep are the subject of much farmer affection. Bec, a Tasmanian sheep farmer preserving the Elliottdale breed on Tommeginne Country, told me who was the 'boss', the 'sentinel' and the 'union leader' within her flock. She went on to describe some of her favourites:

> Phoenix is a character and my bestie. He's a wether [castrated male]. He has had a few health challenges and

> spent a lot of time at the vets…He is halter trained, does an awesome recall and does some tricks…Phoenix pretty much accompanies me wherever I go…He gives the best greetings when I come home after work trips.

Phoenix, Bec told me, is like a 'fun, but exasperated, uncle' to the four poddy lambs they've ended up rearing:

> We started off the year with a 'no inside lamb' rule. Then Wilbur's leg got broken, and he moved in. Norma and Ephram followed when Edith got sick, and they were just so sad outside…I caved and they came inside at 1 am that

Elliottdale courtesy of Sue Curliss

> night…[Phoenix will] discipline them if needed, but woe betide any other sheep who comes near them and tries to touch them. He also taught them where the best treats are kept and how to headbutt the door to be let inside.

In contrast to the focus on averages and uniformity within intensive livestock systems, heritage breed farmers' attention to animal individuality builds connection, while enabling farmers to recognise and respond to any changes in their condition. (Farmers tell me that prey animals tend to mask injuries and illness as a defence mechanism, so it can be difficult to discern poor health if you're not paying close attention.)

The heightened level of care given to individuals is also the result of the scale of heritage breed farming, where the smaller herds and flocks compared with the large commercial operations enable greater familiarity. The endangered status of most heritage breeds, particularly when they're at as great a risk as the carpet-wool breeds, further contributes to the sense that every individual is special. Just as is the case for Bec, each and every one of Sue's Elliottdales is treasured. On her farm Facebook page, she commemorated the birthday of Dottie by recounting the story of her birth:

> Twelve months ago, I found you in the yard on a cold, wet, windy night. I nearly passed you by as you just looked like a piece of loose wool. Your mum took off as there was nothing she could do to help you. I picked you up in one hand and brought you inside. You were so cold your temperature did not register for more than two hours as I slowly warmed you, while keeping your brain alive by rubbing glucose powder on your gums. You did not give

> up and there was no sleep that night as I wasn't giving up either.
>
> Slowly, slowly your temperature came up and I was eventually able to give you a bottle of much needed colostrum, which you happily sucked, much to my elation. Just maybe, you were going to make it. We knew you would have some battles, but you were a fighter, and we would fight with you. Within twenty-four hours, you were standing, albeit wobbly and soon you were walking. Coco looked after you and became your best friend and guardian and from that first night you have flourished.
>
> You had many people cheering you on, Dottie, and in a time of uncertainty and fear, you have been a woolly beacon that we can all get through the bad times if we have friends who have our backs, and we don't give up. It has been an absolute joy being your human mum, and despite our garden being pruned to Dottie height, we love you to bits. Happy Birthday, Princess Dottie.

It is abundantly clear how valued Dottie is by Sue and her social media admirers, for whom she certainly was a 'woolly beacon' during the Covid years. Lamb mortality is a significant issue for sheep farmers. Consequently, intervening in such instances, particularly with rare breeds, is not uncommon, and these poddy lambs often take up a very special place in the hearts of breeders, as much as they bemoan the huge amount of work, mess and lost sleep hand-raising lambs entails.

Love for sheep can also be heightened when they embody links to family history. Intergenerational farmers describe memories of herding or shearing with now-deceased grandparents or parents. In turn, animals themselves are felt to be family. Of his beloved English Leicester sheep, Brenton explained:

> I consider them to be family; they have been our family for over 150 years. I talk to them, and the rams in particular talk to me. Sorry if I sound like a silly old man, but you must talk to them. I gave myself a sixtieth birthday present by commissioning a large portrait of an English Leicester head, which hangs in our kitchen (I do not have a painting of my wife). I have photos, ribbons and certificates of our sheep going back over 100 years, probably more photos of sheep than human ancestors. I can trace the pedigrees of my English Leicesters back for perhaps 80 to 100 years.

I stayed a couple of nights with Brenton and Anne, his wife (who he clearly loves dearly, despite the inequity of portraiture). The next two generations of their family also live on the farm. Anne and Brenton's grandkids showed me their silkworms, puppies, kittens and chooks. On most heritage breed farms, similarly, there is a wide variety of animals of all sorts inside and out. While animal rights activists would have you believe that livestock farmers are callous killers, on the contrary, they are often absolute softies when it comes to animals.

Farmers tend to end up with all manner of rescued creatures in their care. Back in WA, at the Brunswick Show, my daughter and I spent an hour or so looking after two orphaned joeys, while the farmers who had rescued them were showing their sheep. They told me they had got a bit of a name in the region for looking after stray animals, so these joeys had come to them from kangaroo shooters, who had found them in a shot mother's pouch. They initially have to feed them six times a day, including a 10.30 pm and 4.30 am feed, so it's a lot of work.

The walls of farmers' homes are invariably covered in cattle, sheep, pig and poultry photos and paintings, as well as tea towels, clocks, ornaments and curios of all sorts of domestic animals. Bookshelves tend to house a range of livestock-, animal- and flora-related books. In sum, most live and breathe animals and are deeply emotively engaged.

Sheep and Agricultural Shows

One of the primary gathering sites for heritage sheep breeders is agricultural shows. While still popular today, the 'proper' country shows feel like a portal into the past. They illustrate the cultural heritage role of livestock, replete with more than a hint of nostalgia. So, in this final section of the chapter, I want to paint you a picture of my absolute favourite of the many shows I attended in the course of this project: WA's humble but wonderful Dinninup Show.

Driving through the rolling, green hills of the Ferguson Valley, WA's characteristic wide blue skies and red earth contrast vividly with the verdant paddocks. The smell of smoke tinges the air, as farmers take the opportunity to burn off in some cooler weather. It is cattle country, and I pass herds of Angus and Holsteins. The occasional horse is grazing, tail flicking at the persistent flies. There are far more animals to be seen than people, although I pass a few farmers out on tractors ploughing or planting.

As I head further east, farmland is interspersed with tracts of commercial blue gums for logging, alongside patches of native forest. The bucolic feel of the landscape in this beautiful early morning is tempered by the occasional handmade cross, marking the death of a loved one in a road accident. Roadkill kangaroos and the occasional fox decompose at the road's edge. The fires of recent years are written into the landscape, with a layer of fresh green shrubbery juxtaposed against the blackened trunks of eucalypts. The occasional sign advertises hay for sale as well as spray-free apples, pears, pumpkins and plums.

I arrive in Dinninup, on Kaniyang Country, a showground site with—somewhat unusually—no town surrounding it. The oval is already filled with kids playing old-fashioned games, including sheaf tossing and three-legged and egg-and-spoon races. There aren't the big rides of the city shows, but the kids are excited about the jumping castle. A coffee van is anomalous (but welcome!) among the more traditional Devonshire tea stalls, and those selling handcrafted wooden toys, and knitted clothing and dolls.

The poultry show is well subscribed, and the familiar cacophony of galline and human chatter surrounds the displays. I am particularly struck by an ingenious DIY poultry-transporting fit-out on a ute parked nearby. I spend a happy ten minutes tasting a variety of heritage apples, whose enthusiastic purveyors are selling tree stock, but only after checking the location of prospective buyers to ensure the suitability of climate. I am keen to purchase some apples for my daughter, but they are not selling fruit, just giving it away. I wander off, weighed down with apples, touched by their generosity.

But I am here for the sheep. The spinning display is a family affair, where dad does the spinning, and mum and the kids knit beanies, scarves and socks. I buy a beautiful handmade English Leicester beanie for just $18. Close by, the shearing shed is a hive of activity. Shearing competitions by age and grade lay bare how much skill and experience is involved. The younger, less experienced entrants are substantially slower, and the sheep tend to come away with a few more cuts. The roustabouts are all women, who busily sweep and sort the wool. Beside the shearing sheds, the heritage sheep display has several breeds: the magnificent long wool Lincolns and English Leicesters, alongside Shroppies and a couple of Dorset Horns.

I settle in with the sheep people on camping chairs, where they happily sit and talk sheep all day, occasionally interacting with people coming to look at the displays, or heading off to get a coffee, scone or sausage sandwich. I have a lovely chat with a breeder new to sheep, who tells me enthusiastically all about her first lambing season. She says she never imagined you could love a sheep, but she has grown to love every one of hers so very much. She says she and her girlfriend, who she introduces me to later, come to ag shows so they can talk sheep all day away from their husbands, who quickly tire of hearing about them.

Donald, the current president of the WA Branch of the Australian Stud Sheep Breeders Association, tells me that in the old days, the Dinninup Show lasted three days (now it is just one). Families from across the region would drove their best animals on foot for showing. The show's dance was the main social event of the year, and if you didn't find a wife at the dance, you'd have to wait another year until the next one. He talks about the prominent farming families, who still live all over the district. The show holds a great deal of significance for his family. 'It is our traditional show', he tells me. 'My great-grandparents were at the initiating meeting in 1918. The family have been involved ever since.' For eighty-six consecutive years, the women in Donald's family have entered the baking or floristry competitions at the show.

Local families, including several Seventh Day Adventists, established the Dinninup Show to celebrate the end of World War I. Because Saturday was a religious day, unlike most shows, they opted to hold it on a Tuesday; a tradition that continues each year on the first Tuesday of November, on Melbourne Cup Day. The makeshift bar cordoned off by rope is a locus of energy, with the race attracting a great deal of attention in Dinninup.

I have crossed paths with Donald at many a show this past few years. He is in demand as a judge, and I have learned a great deal from him. He talks a lot about conformation, and in helping young judges learn their craft, he reminds them to consider what he calls the Five Ts: type (i.e. true to the breed standard), toes, teeth, teats and testicles. When Donald judges, he places his two hands on the sheep's shoulders before moving to grip every few inches down the length of the body. He is feeling for fat and muscle and says if the skin moves a lot, there is too much fat.

Grant is another respected judge, who similarly focuses on utility traits. At the Brunswick Show some months later, Grant gives the ribbon to the Lincoln with the larger conformation. He explains this by saying that the long wool sheep are today of more value for their meat than wool, so he gives the prize to the ewe 'with the extra chop on her'. Of showing, Grant says that you are the only familiar thing to that sheep at the show, so if you treat them like 'a mate, not a mongrel', the sheep will be calm and good to you. He emphasises that you should never take a non-halter-trained sheep to a show. It will embarrass, if not hurt, you. Grant says you can halter train by tying two recalcitrant sheep to each other. For sheep handling more generally, he is all about keeping things calm. He talks about tying up your dogs if they're stressing the sheep out. And if you're getting stressed, then go tie yourself up.

You've likely got an inkling from Tom's story in the beginning of this chapter that sheepdogs are a huge part of life for heritage breed sheep people. I've bleated on too much about other aspects of ovine life to have space to do them justice here. But we can at least wind up, like the Dinninup Show, by paying homage to these amazingly intelligent and skilled creatures. With the band and the buzz from the bar ramping up as the sun hotfoots it to the horizon, the show's annual highlight commences: the much-anticipated canine high jump.

The crowd of showgoers gathers around a ute fitted out with a metal frame, which houses the wooden slats that incrementally increase the height for the gravity-defying doggos to jump. There is much 'ooh-ing' and 'aah-ing', banter and laughter as the dogs of varying athleticism tackle the task. Just three dogs make it over the 180-centimetre challenge. The suspense is building and, at 2 metres, just one clears the hurdle. Then this extraordinary hound sails over 210...then 220! The crowd is in raptures and even the dog's reserved young female owner cracks a smile. This amazing animal then positively floats over 230 and only ends her extraordinary run at 240 centimetres. A new show record! The crowd erupts!

But every kid grows up knowing that when the dog high jump is over, it's time to go home, so disperse, with cheeks sore from grinning, we then reluctantly do.

Conclusion

When we think about animals that symbolise contemporary Australia, it's quite rightly the kangaroo, emu and koala that spring to mind. But the humble sheep has had a profound impact on the nation's modern history, as ambivalent as this has been. Sheep were at the forefront of the colonial frontier. The profound loss of life, land and livelihoods this engendered for First Nations people, and the ecological degradation sheep caused, cannot be overstated. Yet, settlers justified their presence through a frenzy of agricultural activity and relished the prodigious wealth the wool industry brought. The fine wool of the Merino underpinned the industry, as it continues to do to this day. A corollary of the Merino's popularity has been the decline of breed diversity, which has been further hastened by the rise of synthetic materials.

Yet, the particular qualities of heritage sheep breeds are enormously valued by their farmers, whose love for their breeds of choice is the primary factor that stands between them and extinction. These high-stakes relationships have grown out of generations of complex entanglements between human, ovine and canine worlds. We will come full circle now and end this chapter, as we began it, with Tom, whose words evocatively capture the impact such relationships have in shaping farmer ways of being:

> I think by living a life where there are consequences from forces that you can and cannot control, that directly affect you and animals you care about, you become grounded so very deeply in reality. My son is incredibly empathetic. He cried when his big sister, while playing Minecraft, left

a pixel dog in a cave, she had to go back and get it. He was still upset that evening.

When you are cold, wet, tired, it's getting dark, the animal is birthing, hurt, lost, sick or being aggressive, your dog, with you always, is happy, scared, protective, loyal, hard working. It's deep. It's even deeper when that is your childhood. I have never been more alive than when reality, adversity, empathy, grief and loyalty have flowed with the unwavering setting of the sun and the emotionless cycles of the seasons and weather. Those who have only lived that in their lifetime don't even know they were, or are, more alive, or have been more alive, than those that have not lived that way. I think this is part of what makes many farmers different.

You absolutely can measure a person's life in the epochs of different dogs.

Chapter 3

Pigs

The Large Black pig was first imported to Australia from the United Kingdom in the early 1900s. As hardy grazing pigs subsisting on windfallen fruit and by-products from dairying, they were economical to keep and produced highly valued lard and meat. Since the 1960s, the exponential growth of intensive, profit-driven farming has led to the dominance of hybrids of the Large White, Landrace and, later, Duroc breeds, who have been selectively bred for performance traits such as fast growth, prolificacy and feed efficiency. Large Blacks, along with Tamworths, Berkshire and Wessex Saddlebacks, soon lost popularity, as commercial pig farmers switched to the new model of indoor farming, which promised greater efficiencies and higher yields from the productive white hybrid pigs.

The Gloucester Old Spot, Poland China, Middle White and Welsh pigs are all now extinct in Australia, their bloodlines lost forever. These song lyrics written by Wendy Cope poignantly evoke the extinction crisis.

In Dorset in the days of old
There lived a pig whose hide was gold—

Friendly, beautiful and charming,
Unsuitable for modern farming.
It can't be helped. The world moved on
And all the golden pigs are gone.

For the remaining breeds, the future is uncertain. Pigs are the species in the worst shape when it comes to breed monitoring and preservation. There are now less than 167 registered purebred Large Black breeding sows in Australia, and even fewer registered purebred Tamworths, Hampshires and Durocs. None of the eight recognised breeds exceeded 300 registrations in the most recent RBTA census. Owing to biosecurity risks, the importation of live pigs, semen or embryos is prohibited. As a closed genetic herd, then, preserving existing populations is imperative to maintaining porcine breed diversity.

A major challenge with pig breed conservation efforts in Australia is that very few pigs are being registered. This makes it impossible to know exactly what is out there. Breeders' reasons for not registering are manifold. One producer of heritage breed and commercial hybrid pigs told me:

> I'll be honest with you, I've registered nothing, because it's just a joke. You put registrations in, and it takes 12 months for them to get approved...Once that's fixed, I'm happy to register everything. But, yes, the fees are too dear to be able to register everything. Until we can change that, and yes, get it up and running a bit better, then you might find some of the breeds aren't as rare as you think they might be.

Not all share this opinion, however. Others believe the cost is manageable and see registration as invaluable for breed conservation.

A Brief History of Pigs

Taking a step back, the domestic pig *(Sus scrofa)* has shared at least 8000 years of relations with humans. Most of the world's domestic pigs today are descended from crossing the fatter and more prolific Chinese pigs with the English breeds. Pigs were introduced to Australia via the First Fleet in 1788. Pigs were permitted to roam freely and soon became a major nuisance in the Sydney Cove settlement, as they ate crops, raided food stores and made a general mess. In terms of breed, the Berkshire, Tamworth and Yorkshire were popular.

Until about 1900, when the hunting of feral pigs was increasingly practised, historians suggest that First Nations people did not associate much with this introduced species, with no term for 'pig' known across the many Indigenous languages. Many of the pigs roaming settlements eventually went bush, and today there are millions of feral pigs living across Australia. They are seen as a destructive pest, detrimental to native ecosystems and agricultural lands, and a disease risk. They are despised by farmers and conservationists alike.

Until the 1970s, domestic pigs tended to be kept in mixed family farms, often integrated into dairy and crop farms where they consumed the by-products. Murray, the Red Poll breeder in his seventies who we met in Chapter 1, talked about growing up with pigs on his family farm. Everything the pigs were fed came from the property. They would eat crushed barley, domestic scraps, surplus milk and dead sheep.

Pigs are omnivores and relish meat; however, feeding pigs untreated meat is illegal today given swill feeding increases disease transmission risks.

Murray went on to tell me about the huge changes to the pig itself since this time: 'When we were doing pigs, you only had eight or ten teats on a sow. We'd always try and select a sow with twelve teats, if we could get it. They are now up to twenty!' Along with increasing the number of teats, the pig industry has also focused on lowering fat-to-meat ratios and increasing the middle section of the pig, which contains the most valuable cuts. It used to be a rare mutation that led to extra ribs and longer bodies in pigs, but selective breeding means that now pigs have more ribs. After Murray's dad passed away, they eventually decided to move the pigs on and focus on cattle, which was a decision based on how relatively labour intensive pigs are to manage.

A staggering 90% of pigs in Australia today are raised indoors in intensive systems. While this is an economical and efficient means of producing vast amounts of protein quickly, there are animal welfare issues. In sheds, pigs cannot exhibit their full range of behaviours. Pigs love to root with their snouts, build nests and wallow in mud and water, none of which they can do in sheds.

For workers, life in the pig sheds can be physically demanding and is constituted of repetitive tasks. Consequently, the industry, at times, struggles to find sufficient workers. For both workers and neighbours of large piggeries, the olfactory and physical pollution can be a problem. Groundwater contamination needs to be carefully managed, as does the perennial issue of air pollution.

I have spent time on many heritage breed pig farms, where pastured pigs produce very little smell. Commercial piggeries, however, often emit a terrible stench. In the past few years, there have been instances of Australian piggeries being fined for the intensity of smells that neighbours endure up to 10 kilometres away. Untreated pig waste, as well as poorly managed carcasses, are usually responsible for the odours. That said, a lot of work has been done in the industry in recent years to address these challenges.

The industry has also worked on improving ovulation rates to increase litter sizes, and their success has been phenomenal. In 1990, ten piglets was the average commercial litter, while today some farms have sows producing averages of a whopping fifteen piglets. The sow's capacity to supply nutrients in utero cannot always keep up, however. So along with nutrient deficiencies for the piglets, this can result in higher stillbirth rates, increased runting and weak sows. Piglets tend to exhibit teat fidelity, attaching and drinking from the same teat every time. The average sow has twelve to fourteen teats, so when there's more piglets than teats, piglets often have to be fostered or hand fed.

These practices are pursued in contrast to the breeding priorities of heritage pig producers. Over the last 24 years, Katy, who lives on Dja Dja Wurrung Country, has bred all eight pure breeds of pig. Farrowing, growing and processing them has given her a unique perspective on the need to preserve genetic diversity given the varied traits and qualities the different breeds possess. She describes little interest from the industry in pure breed conservation, and the struggle of retaining the true types given the limited available genetics. She served for many years as the pig coordinator of the RBTA and as Victorian branch president for the Australian Pig Breeding Association in her efforts to support breed conservation.

Tamworths courtesy of Katy Brown

Katy describes her approach as one that values the breed's adherence to the standards of excellence, rather than the desire to breed the optimal commercial cross based on terminal traits:

> One of my strictest criteria for breeding is soundness and the ability to mate and birth naturally, to milk and show maternal instincts, and to breed and remain sound into old age. Many old breeds continue breeding well after modern breeds have finished. Soundness of skeletal structure is overlooked in many breeding programs in favour of fast growth and meat deposition. 'They only have to walk to the truck' is a phrase I hate. Longevity is an important factor for productivity, especially when the animal is part

> of a subsistence process where poverty stops people replacing stock. Usually the higher producing, the shorter the useful lifespan, as the animal simply wears out. Welfare is an important part of farming and good selection can mitigate a lot of simple issues like lameness, foot issues, face biting. Good immunity and nutrition are instrumental in maintaining quality and bone in animals.

After raising all the pig breeds available in Australia, the critically endangered Tamworth is her breed of choice. She described them as 'vocal, smart and pretty self-sufficient'. She said while they won't suit everyone, they 'generally raise every pig they birth, come rain, hail or shine. Teat issues are non-existent, and they have terrific feet. They have relatively slim long piglets so rarely need assistance and milk for six to eight weeks without dropping too much weight.'

Radical Pigs

Not far from Katy, also on Dja Dja Warrung Country, Tammi and Stuart own Jonai Farms & Meatsmiths. At a species level, pigs and poultry suffer the poorest animal welfare outcomes given the widespread adoption of intensive systems. This led Tammi and Stuart to their decision to raise pigs. When deciding on which type of pig, they looked to endangered breeds, as it was important to them to contribute to maintaining resilience in agricultural systems by supporting breed diversity. They opted for Large Blacks, as they are among the most endangered breeds, while also being good eating, great mothers and docile in temperament, which was essential given they had young children when they started out.

Tammi and her family then added cattle, including Dairy Shorthorns, to the mix, which are well suited to their hilly, volcanic country. Of her reasons for raising pigs, Tammi commented:

> Farming cattle is like a shitload easier than pigs. And there's a part of me that would just abandon the pigs and grow cattle and make it a little easier on ourselves. But we need more of us to be growing pigs properly. So we're committed to that.

Along with being a farmer and butcher, Tammi is a committed activist, the long-serving president of the Australian Food Sovereignty Alliance (AFSA), and a scholar, who recently completed her PhD as part of this heritage breeds project. Her positioning between these worlds makes her family's farm a fascinating case study of agroecology in action, where every decision foregrounds their strong commitment to social and ecological reparations.

On an icy cold morning in July 2023, I joined Tammi, Stuart, their 19-year-old son and a group of ten friends, including neighbours and members of their Community Supported Agriculture (CSA) group, in walking out to the paddock where an 18-month-old Large Black barrow (castrated male) was grazing. While Tammi talked her son through where to make the cut on the pig's throat, Stuart used food to tempt the pig onto the grass and away from the mud so the carcass would stay relatively clean. The pig appeared totally relaxed as it ate from the bucket of feed Stuart held. Once in position, from close range, he shot the pig, killing him instantly. Four people rushed in to help lift the pig, as their son slit his throat. After the pig was bled,

Tammi rapidly whisked the blood she had collected for the *morcilla* (blood sausage) in a large bowl to prevent it from coagulating.

The carcass was then carried by tractor to an outdoor bathtub. Water was boiling over a fire in a large vat nearby, and we formed a line and passed along buckets of boiling water to fill the bath. The pig was then placed into the scalding bath, where we commenced scraping its thick, black, coarse hair off the skin. With four people at a time working away at the hair with various scraping tools and knives, the pig's white flesh was soon exposed.

After the cleaning, the pig was gutted, with the offal saved for eating. Tammi cleaned the stomach to house the *morcilla*, the lungs were cooked in a sour sauce, and the heart skewered with chillis from the garden and barbecued, along with the spleen, for our lunch. The liver was reserved for pâté, and the dogs were on hand to snap up any scraps. The next day we made salami. Tammi quartered the pig, reserving one hind leg, which she cured with rock salt, while the rest of the meat was chopped up and made into various types of salami.

During *la Matanza*, the annual killing and processing of a pig for domestic consumption, Tammi gives thanks to the pig for nourishing her family and community over the year ahead. Tammi is pragmatic about killing her animals, though it's not something she enjoys. She is an animal lover of old and was a vegetarian for many years. A Berkshire farmer captured the sentiments of many farmers in terms of the relationship of reciprocity underpinning feeding and feeding upon:

> From the moment they're born, I love my pigs, and they love me. I make every meal for them. I put them on fresh

pasture. I make sure they have mud to wallow in and water to drink. And, at the end of the day, they make sure me and my family have food on the table and love in our hearts.

La Matanza gives a taste of life at Jonai Farms & Meatsmiths, where Tammi and her family eschew commodity grain, feeding their pigs entirely on locally available waste-stream feed, including spent brewers' grain, whey and several container loads each year of out-of-date or damaged food, such as strawberries and muesli bars. Collecting and feeding waste-stream feed otherwise destined for landfill is time consuming and labour intensive, but an important means of minimising their ecological impact. It is also part of reducing their reliance on expensive farm inputs and outputs, as is Tammi's practice of doing the farm's butchery.

Determined to be as self-sufficient as possible, in 2013, Tammi and Stuart crowdfunded and built a butcher's shop and commercial kitchen, where they produce a range of fresh cuts, smallgoods and charcuterie. Surplus fat is made into soap, pigs' heads into pâté, and bones into broth. Bones that remain after cooking are pyrolised to create bone char. This is then activated in barrels of biofertiliser made microbially rich with fresh manure contributed by their house cows. This, in turn, is used to nourish a small commercial crop of garlic and a diverse range of heritage-variety vegetables to feed the farm community. Tammi and Stuart are close to reaching self-sufficiency in food, where, along with pork and beef, they receive milk from their beloved house cow, Wynny. They transform a portion of this into a range of cheeses, along with growing vegetables and heritage-variety wheat for the household breadmaking.

Tammi has recently trained to become a licensed meat inspector in anticipation of building a micro-abattoir, which they now have approval for on their farm. This will reduce stress for animals, support other local heritage breed farmers struggling to access facilities, and it closes the loop in terms of reliance on external services.

Most of the farm's produce is sold through their CSA network. They deliver a monthly bag of mixed fresh and cured meats to eighty households, many of whom have been with them over the ten years they have operated as a CSA. CSA members supported Jonai Farms through the horrendous loss of the majority of their cattle to acidosis in January 2020, and Jonai Farms, in turn, supported members through the Covid pandemic, getting food to them while they endured 5-kilometre limits on movement during lockdowns, with special treats like jars of sourdough starter and Vietnamese *phó* broth, along with a collection of Tammi's recipes.

Tammi and Stuart work hard to be good custodians of the land. 'We start with the soil and work up from there', Tammi told me. 'We try to have 90 to 95% ground cover at all times, which is a challenge with pigs, especially in winter when they're turning the land more quickly. So higher-paced rotations are really important to our system.' Even though they have a long waiting list to join their CSA, they have a no-growth model, and ensure they respect the land's carrying capacity by not overstocking. They see the benefits of pigs in fertilising their paddocks. Of pig manure, another heritage breed pig farmer told me: 'It's bloody good stuff. It's hard to beat pig shit, actually.' Pigs' rooting behaviour can also play a positive role in regenerating land. They are particularly good, I am told, at getting invasive plants like tobacco and lantana under control.

As non-Indigenous Australians, Jonai Farms share the benefits of their land use with First Peoples by paying 1% of their income to a local First Nations organisation, Pay the Rent. In addition, in their efforts to progress social justice, since 2022, they have shared part of their land in a rent-free arrangement with Josh, a Ngarrindjeri and Narunnga man, and his settler farming partner, who operate a market garden, Tumpinyeri Growers, on the farm. Through her work with AFSA, Tammi is constantly looking for ways to progress social as well as ecological justice. This cannot be done alone, and she is always focused on collectivising to work with others to these ends, and to address the challenges inherent to pig farming in the current economic and legislative environment.

Challenges Raising Pigs

Small-scale heritage pig farmers face many hurdles. The cheap pork enabled by the intensification of the industry has accustomed consumers to paying less for pork than it costs small-scale farmers to produce it. Farmers also have to navigate a complex regulatory environment around pig keeping. Vets specialising in pigs are few and far between and vaccines are sold in bulk, as they're marketed to large-scale producers, making them expensive.

Pigs are also more labour intensive than grazing animals to keep, as they require exogenous feed. Additionally, they need to be regularly moved in order to rest land impacted by their rooting behaviours. Pigs also require careful management to cope with extreme weather. Farmers pay close attention to nutrition in the cold and provide increased feed and bedding materials. Heat is even trickier, as pigs don't have sweat glands (ironically, given the erroneous adage 'sweat like a pig'), so need to regulate their temperature by wallowing in mud and water during hot weather. In a heatwave, Katy described doing the rounds every few hours to ensure her pigs had enough water. Katy recounts the struggle of sows labouring to give birth in the intense heat, and how she gives them extra water with glucose and electrolytes, staying with them during the whole birthing process, often through the night.

According to the Australian Bureau of Statistics, farmers work longer hours than most of the population, putting in an average of 57 hours a week, compared with the national average of 44 hours for full-time workers. People with pigs and dairies work hardest of all. Despite the hard work, making ends meet has become ever more difficult for pig producers. For those buying in feed, the war in Ukraine along with inflation have increased costs considerably. Global competition also makes life difficult for Australian pig producers. While all fresh pork in Australia derives from pigs raised domestically, around 80% of our bacon and ham is imported, and local producers just can't compete with the low prices of international producers.

One of the biggest challenges small-scale livestock farmers currently face is accessing abattoirs and processing infrastructure. Countless small regional abattoirs have shut down as the industry has centralised, with in-house slaughter part of the vertical integration model of major pig producers. I talked about this with a butcher in WA:

> If you decide to grow an amazing chicken right now, you're fucked. Where are you going to get it killed? When I was

> younger, there were 10,000 abattoirs, and now, there's nothing. And too many rules as well, and stuff where it's the overkill of protecting the public. How many people have died from eating meat? And yet the rules are so strong. Why shouldn't a farmer be allowed to do a course and get a certificate for slaughtering? But he shouldn't have to build an abattoir. It should be just a simple shed, with a good cold room and a good boning room.

Many farmers have to drive huge distances to reach abattoirs that will take their relatively small number of animals. The costs in time and fuel are substantial, and animals suffer the stress of the long trip. Moreover, heritage breed farmers often lose valuable meat because the abattoirs are automated. Mechanisation doubles the speed of the line, but comes at a cost for those with non-standard sized animals. A further complication for pig producers is that halal-accredited facilities won't transport or kill pigs.

Disease in Pigs

Alongside the challenges already outlined, pigs are susceptible to a range of diseases. Risks are exacerbated by our globalised economy and changing climate, and zoonotic diseases have ripe grounds for catastrophic outbreaks among genetically similar animals raised in conditions of intensive animal confinement. Katy explained the unique positioning of pigs in relation to zoonoses:

> The danger with pigs is that they are a wonderful multiplier of viruses, and the great fear is that, say they got exposed to an avian flu and a human one, it may mutate into a

zoonotic disease. And pigs are so like humans. This is why biosecurity around pigs is so important.

With African swine fever on Australia's doorstep, and there already being unprecedented outbreaks of Japanese encephalitis (JEV) in the southern states, the risks to pigs posed by virulent disease are increasing. For those with endangered heritage breeds, disease outbreaks are particularly concerning, as whole bloodlines can be lost.

African swine fever is a highly contagious virus, with a morbidity rate ranging between 80% and 100% in pigs. (Humans, fortunately, are not susceptible.) There are no vaccines or treatments. Since 2007, the disease has spread throughout the world. Our biggest trading partner, China, has had successive outbreaks since 2018. Outbreaks have followed in Indonesia, Timor-Leste and Papua New Guinea, fuelling fears that the disease will soon reach Australia. This is because the virus survives readily on material such as clothing, shoes and tyres, as well as in pork products, like ham and bacon. Biosecurity measures at international borders were ramped up substantially after the disease hit Bali in 2020. While many goods containing the virus have been detected in airports, so far the disease has not infected pigs in Australia, but pig breeders remain very nervous about this threat.

We haven't been so lucky with Japanese encephalitis. In February 2022, JEV was detected in a human for the first time in NSW. Before this, only one person in Australia had ever been diagnosed, and that was in north Queensland in 1998. For this tropical disease to be so far south was unprecedented, and a worrying expansion of JEV's range. Courtesy of the La Niña weather pattern, November 2021 was NSW's wettest month since record-keeping began in 1900, and the rain kept coming all summer.

Within a week, JEV had spread to Victoria, Queensland and South Australia. Mosquitoes, the disease's transmitter, multiplied in staggering swarms, as flooding impacted multiple regions. Katy described the mosquitoes as rampant: 'I'm in a flood area, and I swear the mozzies are hatching by the second, the back fence is sagging under the weight of them.'

Of the forty-five people who were diagnosed and fell ill with the disease, six lost their lives in the outbreak. Those who survive often struggle with ongoing challenges, including cognitive impairment, convulsions and general weakness. While there are safe and effective vaccines, which the government was quick to roll out to those at risk, such as piggery workers, there are no approved treatments.

We cannot be sure exactly what happened, but experts posit that JEV likely arrived via infected migrating waterbirds, who flocked to the Murray–Darling basin during the wet 2021 season. The swarms of mosquitoes meant that the virus soon spread to high-density piggeries, which provided rich grounds for the disease to flourish. More than eighty piggeries ended up infected with JEV. In pigs, JEV causes sows to birth stillborn and deformed piglets, with whole litters often lost. Katy says that detection was slow:

> The initial outbreak in NSW affected quite a few, but it was over before anyone knew what it was. It was here [in Victoria] at least six months before detection, and it was really the deformities that triggered the vets to look outside our normal range of bugs. In initial cases, people thought [it was] lepto or even PRRS [porcine reproductive and respiratory syndrome].

The costs were high for pig farmers, and many spent thousands of dollars on mosquito control, including parasiticidal drenches. A Wessex Saddleback farmer in NSW lost only one litter to the disease, but he described the ramifications as far wider reaching. His property was subject to a control order, which meant he couldn't sell or move pigs for six months. He consequently had no income from his pigs during this time, while bearing the cost of feeding the fifty-odd piglets born before the outbreak that he normally would have sold. He has since done a lot of work to control for mosquitoes, putting in drainage systems and using gravel to dry up areas of pooling water.

More than the financial and practical challenges, the most unpleasant aspect of the experience was the stigma he felt from his local community. Anxieties were already heightened around disease because of Covid, so people were wary about the health risks posed by the outbreak in his pigs. The mental health stress experienced by impacted farmers was great. The pig farming community is supportive, though, and Katy was proactive in bringing this issue to the fore through a JEV Facebook page she started, where she posted the following:

> I just want to talk about mental health when things are so uncertain. If you are like me, you will have a feeling of dread that it's not if, but when...It is with a sense of dread that I checked my due-to-farrow sows this morning, and I think this is going to be the way it is. This hypervigilance is very stressful for your body long-term, so I am urging you all to take care of yourselves, and if anyone needs any support, reach out.

With the increase of new disease risks, biosecurity is of growing concern. During the outbreak of African swine fever in Bali, I attended a government biosecurity webinar targeted at farmers. Experts say Australia has world-class border security, but much less systematic biosecurity at the farm level. Moreover, historically, in pigs, we have experienced many more bacterial than viral threats in Australia, so most of our protocols are designed around protecting against the former.

Among heritage breed farmers, there are varying levels of anxiety and action around biosecurity. Some pig breeders had no biosecurity protocols in place for visitors; some asked that I avoid visiting other pig farms on the day I came to theirs; some kept me some metres away from their pigs; some had me wear protective booties; and one had me change into their overalls and boots to visit the pigs. I had an interstate trip booked to visit a series of farms when Covid first emerged. I wrote to one pig farmer I had planned to visit asking if she thought I should cancel. Her reply was quick, and she said she was more concerned about pig disease transmission than Covid. I never made it to her farm, unfortunately, but her care for her pigs above herself was notable.

Along with these relatively new viral threats, there are other common diseases that pigs are susceptible to. During a conversation between members of the RBTA Board, Judy mentioned that one of her five-week-old pigs mysteriously died: 'No idea why he died, unless a snake bit him, but poor old snakes, we blame them for everything don't we, so I really don't know.' Katy replied that it sounded like GPFD. It transpires this is a common ailment, where cause of death remains indeterminate, that is: Good Pig Found Dead.

Loving Pigs

So why do people persist in raising pigs when there are so many challenges? Pigs are highly charismatic, animated beings, who are dearly loved by their champions. Christine, who has long been a passionate promoter of Large Black pigs and lives on Dja Dja Wurrung Country, showed me five of her Large Black sows who weren't pregnant because she didn't have the space or the market for more piglets. Yet, she is keeping the sows, not knowing if they'll be too old to conceive again when she will next need more piglets. The cost of feeding them is not insignificant, and even though finances are tight, she can't bear to part with them as she loves them. This is not least because Large Blacks have impressive longevity, which means relationships build over the years. I met one of Christine's favourite Large Blacks, Priscilla, who is still having healthy litters of piglets at 14 years.

Always entertaining to watch, pigs get so excited about being fed and devour their food with gusto. They are very noisy creatures, and along with squealing at high decibels, they have a broad range of vocalisations. They have a heightened sense of smell, hence being used for finding truffles.

A pig's genetic makeup is close to our own. In some cases, their stem cells and organs are used for medical treatment in humans. And pigs are incredibly smart. Tammi talks about their pigs outsmarting them: 'So we've got a Duroc boar in the herd now, Phoenix. And he jumped so many fences in the early days. He completely rewrote our breeding program by himself.' Katy similarly told me about pigs' unique qualities:

> Pigs are undoubtedly the most intelligent of the farm animals and are one of few that make direct eye contact. Pigs are highly motivated by food and thermal regulation;

Large Black pigs courtesy of Tammi Jonas

> hot pigs will leave litters, have litters in water etc, the driving force is survival. Pigs are similar to dogs to work with and are easily trained. They can be quite interactive e.g. [my son] when he was about five, had a hand-raised gilt called Piddle. They used to make mud pies, she would push the mud up, he would scoop it out, over and over. She would also cover us with straw if we lay in her bed. She knew several vocal commands and would happily go for walks like a dog. She is one of many pet pigs we have had that had quirks that would be thought of as showing personality.

As with all animals, temperament varies between breeds and individuals. Many of the heritage breeders in Australia are women, and Katy suggests that temperament is usually good in pigs because women won't tolerate aggression for safety reasons. While good temperament is valued, pigs can still be aggressive, which can also be a positive in some situations. Tammi told me she 'had a couple of sows kill a fox not long ago. Yeah, took him down. We were like, go you!'

On the whole, though, because they require feed, pigs tend to be in proximity to people daily, and so easily build positive associations with people. I visited a piggery in the Barossa which, along with white commercial crosses, bred Hampshires. When we entered the pens, the piglets ran to us and mobbed us, nibbling at our gumboots. It was hard not to fall in love with them.

Meat and Fat

As we have seen with other livestock species, loving animals does not preclude eating them. Pork is a rich, flavourful meat that is the most consumed globally alongside chicken. I have heard more than one reformed vegetarian say that it was bacon that broke their resolve. As is the case for all species, different breeds of pigs have varying qualities, and this is true of the meat profiles of each breed. Tammi explained:

> I have butchered a lot of animals for other people, so I've got to see differences in breed and age, feed, and all of those things. There's an undeniable role of breed in things like whether or not their meat will marble. It's just not a common thing in any of the industrial breeds, and you'll see it more in certain, only certain, heritage breeds...So Durocs have got bigger hams and loins, and they marble, which we don't get in the Large Black at all. And the piglets are born bigger, and then they grow much faster. The growth rate is much quicker, but, of course, you know, we have patience, and it tastes delicious, and we're not here for yield, we're here for ethics and quality.

Success Stories: Berkshire Pigs

Berkshire pigs are currently enjoying popularity on account of their eating quality. Some of the bigger mainstream piggeries now have Berkshire lines. Not all of their bloodlines exhibit marbling, but many do, and the Berkshires' reputation for high-quality meat means they are coveted by the high-end restaurant market.

The increasing popularity of paddock-to-plate restaurants and lifestyles is also helping. In the midst of my fieldwork in South Australia, I returned from a day at a piggery in the Barossa, and the farmer had taken a photo of me with his Hampshire pigs. He sent the photo to a mutual pig farming friend back in WA telling him I now preferred

Hampshires to Berkshires. While this amusing banter was unfolding, I watched Jamie Oliver on the SBS food channel making sausages out of Large Black and Wessex Saddleback pigs. Next up was *River Cottage* and Paul West was talking about his Berkshires. You could be fooled into thinking that heritage pigs are thriving given how valued they are in this genre of chef/lifestyle programs. Lovers of the Berkshire breed like Paul West say they are hardy animals and excellent mothers, which was reiterated by a legend of the breed from WA, who believes: 'She is a far better mother than any other breed and she is so kind.'

Few would dispute that the Berkshire is a better tasting meat than the white breeds. However, the main deterrent to people eating them is their black hair. Kyla, the butcher we met in Chapter 2, explained the problem:

> In the butcher shops, I had such a hard time with the Berkshires. One, because of the fat on it. So, you take the skin off. But then people go, 'where's the crackle?' But then you leave the crackle on, and it's the black hairs. So, even if you shave the hair, the follicles put people off. But what people buy to eat at home and what they buy out at a restaurant are two very different things. So, they'll have a Berkshire belly or Berkshire chop or whatever...They'll go to a restaurant and eat something like that. They go, 'oh, that's bloody beautiful!' They come in [to the butcher], they go, 'do you have Berkshire?' You go 'yes, it's that'. They go, 'oh, it's a bit fatty, oh, there's black hair'. You're like, 'that's exactly the fricking same as the white hair, all over, that's it!'

All pork has hair, but the pale hair of the Large White/ Landrace/Duroc crosses of the commercial producers are less visible, so less concerning to the consumer. Kyla went on to describe how people will similarly love a marbled beef steak at a restaurant because of the deliciousness the fat endows, but seeing it raw, they will gravitate towards the leaner option.

Fat, then, has also inhibited heritage breed pigs' popularity. In the 1980s, fat was stigmatised as bad for our health. The overly simplistic messaging around fat's properties at the time has since been qualified, but the impact on consumer perceptions has been profound and persistent. Breeding lean pigs quickly became the industry's focus, and today a preference for lean meat continues. I was talking to my local butcher about this (who, incidentally, is the seventh generation of butchers in his family, with his adult son also following in his footsteps). He was telling me that he has problems sometimes with receiving Berkshire pigs that are too fat. This means lots ends up being trimmed, so profits are lost. Tammi also described the challenge of getting fat ratios on their pigs right, especially when their waste-stream feed is quite variable. This is one of the challenges of heritage over commercial pig breeds, but she always finds uses for the fat either in sausages and chorizo, or in making soap.

There is, however, growing recognition of the gastronomic value of fat. More specifically, marbling, or intramuscular fat, is increasingly desired for the tenderness and flavour it brings to meat. Several factors determine the presence of fat location, quality and quantity. It is not only breed and the animal's age, but also feed. And along with its culinary qualities, pig fat is valued for making soap. Some Wessex Saddleback breeders described having a growing market for their lard soaps to the extent that they are growing their pigs out to a greater weight and an older age to intentionally produce more lard.

Breed as Brand?

Given the challenges of economic viability with heritage breed pigs, in this final section we will look at how farmers make ends meet, including probing the question of the value of using breed in branding a product. Farmers use creative means to integrate their heritage breeds into various business models. We have learned already about Tammi's approach with direct sales via a CSA model. Tammi and Stuart also offer regular workshops to share their learnings.

I spent a few days staying at Kim and Daniel's farm on Tommeginne Country in Tasmania, where they primarily grow Berkshire pigs. They told me they had spent time learning with Tammi and had been influenced by her no-growth model. They, too, aim only to sustain themselves and live a quiet, good life, rather than pursuing the dominant model of seeking to grow their business and profits. Relative to their former urban lifestyle in Brisbane, they need less money to support their rural lifestyle. Daniel trained to be a butcher and spent time learning in Tammi's boning room, and they sell a range of pork and charcuterie products at the Hobart and Launceston farmers' markets. They also have a farm shop, where, along with fresh and cured pork, they have a range of prepared meals, including Boston baked beans made smoky with pork ribs, and a whole lot of pickles, jams and other delicious morsels. They also sell to restaurants locally and as far as Hobart. The majority of the farmers

I met, irrespective of the species they farm, relied on at least some off-farm income, so it is heartening that they are managing to support themselves with no off-farm income.

One of the things I have been somewhat surprised by, with the exception of Berkshire, is the relative lack of marketing of meat and other products by their breeds. I went to one incredible paddock-to-plate restaurant in a small rural town in Queensland, for example, where I knew that heritage breed pork from the owner's farm was featured. Yet, there was no mention on the menu of the pigs being rare or heritage breeds. I asked the waiter what breed the pigs were, and she had no idea. She went and asked the chef who advised her they were Wessex Saddleback, Berkshire and Large Black.

While it seems like a missed opportunity to me, perhaps the reason they didn't promote this unique aspect of their offerings was because consumers, for the most part, have very little awareness about livestock breeds. One of Perth's charismatic butchers didn't mince his words when explaining this in relation to cattle:

> It's never been popular. There are two that stand out: Wagyu and Angus. And it has to be Black Angus, because the moment you tell them about Red Angus, huh? I've taken an interest...But that's never hit the public, and the public really couldn't give a fuck.

Yet, Tammi strongly believes it is the farmers' and butchers' role to educate consumers:

> I've never found it remotely insurmountable the complexity of the discussion, but I have found a lot of my other farming comrades do. They, yeah, they get very

> quickly, like 'argh, people don't get it! Now they're all going vegetarian!' Yeah and farmers are kind of coming to it with an old industrial mindset in a way. 'I'm here to grow food, flog it, and you're here to eat it' and that's it. And you know people like me who are devoted to CSA and building different kinds of food economies are like, 'no, I'm here for the conversation. This is a relationship, not a transaction, and my job is to help educate you as an eater about what it looks like to grow food.'

This idea of farming as inherently about relationships rather than transactions is a theme that arose in conversations with several heritage breed farmers. In trying to progress rare breed conservation, Katy is working with the same principle but in a different way. Over her many years of farming pigs, she has observed that new pig farmers tend to last only about two years given how difficult it is. Because of the high attrition rates, she has started approaching conservation differently:

> We have personally sponsored three young people into pig breeding by providing the breeding stock for free and giving them a guaranteed market for the piglets at a good price. This is proving successful, as the hardest part for any new person is finding the capital to start with stud pigs and developing a steady market. That's probably going to be the way forward for us. I have lost faith in selling pigs to people, as have several other long-term breeders, as everybody loves the idea, but once they find it's a seven-day-a-week job, and the market challenges, they bail out and the pigs disappear. This way I can at least protect the breeders.

Conclusion

Where food used to be primarily understood through a lens of human sustenance and sociality, today food is just as often constructed as a commodity where the goal is maximising profitability. Within this paradigm, the shift to corporate ownership of seeds, breeds and bloodlines has occurred alongside the commodification of livestock. Supporting food production practices that support animal welfare, the environment and farmer wellbeing has never been more important. Yet, small-scale pig farmers in Australia are under unprecedented pressure as they battle to make a living in the face of price cutting from large-scale corporate agribusinesses locally and globally, loss of critical infrastructure, dynamic consumer values, climate change and growing biosecurity risks.

Pig farmers and activists like Tammi and Stuart are working hard to collectivise and fight for a more just food system. On their farm, they have resumed control over each part of the process by managing the breeding, raising, butchery and sale of their pigs and cattle themselves. Soon they will have closed the loop through also slaughtering their animals in their own on-farm abattoir.

A growing number of farmers are foregrounding relationships over simplistic transactions in their efforts to care for the land, their animals and human thriving. While economic value is oriented solely around human economic gain, they are reconfiguring ideas around value with an approach that seeks to ensure the wellbeing of all participants in the food chain, from the microbial to the planetary level. Within these kinds of systems, pigs, and indeed other livestock, are loved and valued as sentient companions embedded in complex interspecies interdependencies grounded in relations of reciprocity.

Chapter 4

Poultry

Born in 1954, Jill grew up on a dairy farm on Tommeginne Country in northern Tasmania. Her father introduced her to poultry keeping when she was eight or nine, as he wanted to teach her how to run a business and do book work. She had White Leghorns and dutifully recorded the number of eggs and who she sold them to, while taking care to balance the books. From this early introduction, Jill told me:

> I just loved anything with feathers. I used to go along as a kid to a neighbour down the road. He had racing pigeons, and I used to steal them and bring them home! And of course they'd always fly back to him. He didn't mind. He knew where they were and that they'd come back.

Throughout her life, Jill has kept numerous types of poultry. When her daughter gifted her Japanese Bantams, she started showing at local agricultural shows. This greatly increased her knowledge of the different breeds available and ignited her passion for rare breed conservation. Today, as well as the Japanese Bantams, she has Wyandottes (white and silver pencilled), Phoenixes and New Hampshires, along with Brown Chinese Geese, Guinea Fowl, Pheasants and Elizabeth Ducks. She devotes a considerable amount of time to volunteering in local poultry associations, as well as serving as the poultry coordinator for the RBTA.

Chickens are loved by aficionados like Jill as quirky and practical pets, whose eggs are enjoyed in return for companionship and care, within a long history of human-galline reciprocity. But over the past half-century, many Australians have come to see chickens as merely cheap meat. Chickens today are the most widely distributed and consumed of the domestic species globally. The majority of the world's chickens *(Gallus gallus domesticus)* are commercial meat birds, known as broilers. From only 5 kilograms eaten per Australian household in 1960, chicken consumption has increased exponentially to over 45 kilograms per capita annually today, along with 250 eggs. The exponential rise of chicken and egg production has been enabled by selective breeding for productive traits that has resulted in a cheap, convenient and ubiquitous protein source. Yet, this has come at a cost to chicken welfare.

A Brief History of Chickens

From the Jungle Fowl's origins in Southeast Asia, over 1600 distinct chicken breeds are the result of natural and artificial selection. Interestingly, the domestication of chickens in around 1500 BCE was considerably later than most domestic species. Domestication arose in tandem with rice agriculture, where the year-round presence of grain likely attracted fowl into human worlds. Traders brought chickens to Europe in the first millennium BCE, where archaeological evidence suggests they were kept as exotica for several centuries before becoming a food source. Eating chicken was only popularised in the Anglophone world during the Roman Empire's expansion. Intriguingly, during this period in Britain, chickens were, at times, buried with people, which archaeologists theorise may have either been as guides to the afterlife or as food offerings.

European colonisers introduced chickens to Australia via the First Fleet in 1788. Hens were kept primarily for their eggs, an economic and nutritious protein source, which people have long valued, particularly in times of turmoil. During World War I, the United States government proclaimed keeping backyard poultry to be a patriotic duty, and recommended two hens per household member to keep the family fed. During World War II, the shortage of agricultural workers and a brutal drought led to food shortages and similar campaigns in Australia. Householders were encouraged to grow 'victory gardens' and raise fowl to feed the family.

In stark contrast, today's entrenched cultures of urbanism have resulted in backyard chicken keeping being frowned upon by many local councils. Due to their crowing, rooster keeping is forbidden in most urban areas. This hasn't stopped Ernest, however, an Araucana chook enthusiast in his mid-seventies living in NSW's Central Coast. 'I do keep a rooster', he told me conspiratorially:

> He sleeps in an underground bunker of a night. Now what I do is, I buried a doghouse in the sand at the side of the chook house. It has a ramp down into it...He goes to bed of a night, about 5.30, then he comes out at nine o'clock in the morning.
>
> You wouldn't do that to hens. Hens are out scratching around for food really early...They actually will come

> down outside my bedroom door. And they talk to you in the morning and say, 'give me breakfast!', because I understand the chook talk.
>
> Now, the boys don't need as much food; they're not big eaters, they don't have to produce eggs and stuff. My rooster, he's a cool dude, and he's nonchalant and just moves around. All the animals have their own personalities...
>
> Now, I'm being sexist, Catie, but the girls are much happier with a boy around. He will sort out any squabbles... So, I've found sunshine...excuse the expression, sunshine and sex produces a better egg.

Ernest much prefers his own eggs to buying them. He grew up with various breeds of backyard poultry that were kept for eggs, and only on special occasions would they eat an excess cockerel or a hen past laying age. I ask him to elaborate on the need for sunshine:

> You need a lot of sunshine on your chickens to get the vitamin D. Because when you've got a chook inside a big shed all the time, and they don't get any sunshine on them, and they're laying eggs, I don't think you're going to get enough vitamin D. So, the sunshine is for their own health, they need a lot of sunshine on their feathers.

Ernest takes pride in giving his chooks a happy and healthy life, and his love for them is clear. He lives alone, so they provide not only a sense of purpose but also companionship, including early morning chats and an alarm clock service.

While Ernest's approach to animal keeping may not, at first glance, sound revolutionary, sex and sunshine play little part in the production of many commercial chickens.

Hens are artificially inseminated, with fertile eggs raised in incubators. Chickens are predominantly raised indoors, where they are exposed to up to 16 hours of artificial light a day to stimulate feed consumption and, in turn, consistent eggs for layers and fast growth for broilers.

Broilers

Let's now dig a little deeper into the industrialisation of chickens. A surprisingly small number of trademarked chicken brands completely dominate the markets for broilers and layers respectively. Their genetics are owned by multinational corporations that supply most of the world's chickens. The broiler industry has flourished through its eye-watering efficiencies that have been achieved through intensification and vertical integration. This is in conjunction with just-in-time production and distribution—characterised by the rapid throughput of chickens from incubation to processing to retail shelves—and a responsiveness to consumer desires manifest in the constant diversification of value-added chicken products.

While there's diversity in retail chicken product offerings, with numerous cuts, preparations and flavourings, this does not extend to the birds themselves. Dizzying growth and feed conversion rates are embodied in the Ross and Cobb chicken brands, from whom all commercial chicken meat in Australia derives. (Google them and you will see that they are marketed as 'products' rather than animate beings, with lists of specifications that read more like those of a laptop than an animal.) These genetics are imported from overseas by two companies, which control over 70% of the national market.

From the 1930s to the 1990s, the average size of a broiler increased by 65%. What makes this so astounding is that the time to reach full growth has over the same period been reduced to just five to six weeks, where non-commercial chickens take around four months to reach maturity. Broilers' rapid growth in such a short space of time comes at a cost to their welfare, however, as they live with an insatiable hunger that dominates other natural instincts and behaviours. Leg deformities, trouble with locomotion, pain, laboured breathing and lameness are all well documented in these fast-growing broilers.

A heritage breed pig farmer told me about the issues he has with the commercial broilers he buys and grows out each year. (He raises a few sheep for the freezer, too. He wants to ensure all the meat his family eats comes from animals that have had a decent life, and also that he 'knows what's in them'.) He says that the broilers grow far too quickly for their own good: 'Problem is you lose half of them, because they just grow too quickly. Their organs fail.' He carefully manages his chickens' feed so as to slow their growth and prevent some of these problems. But their behaviours are written into their genetics, so there's limits on the impacts of even the best management. Herein lies the issue: while consumers are deeply concerned about animal welfare, no amount of free-ranging space or enrichment opportunities change the fact that these commercial broilers have been selectively bred to grow at rates that appear to outpace their capacity to lead comfortable lives.

The ready availability of this cheap chicken has transformed Australian diets. It is interesting to observe how, over time, chicken's positioning—not least the result of clever marketing—has reflected changing values and practices in society. Before the 1960s, hens were kept for their eggs, not their meat. Then, in the seventies, chicken meat came to be valued as a fresh, healthy centrepiece of the table, while in the 1980s, it was touted as a low-fat option.

In the 1990s, chicken, in its many value-added, quick-to-cook forms, was praised as a convenience food that liberated overworked mothers from complicated meal prep. Indeed, as a convenience food, the pre-prepared barbecue chicken continues to be somewhat of an institution, which today is gloriously badged as 'the bachelor's handbag'. In the new millennium, chicken is pushed as a more ecologically friendly meat because of fewer greenhouse gas emissions in their production than for ruminants. Responding to concerns raised about the welfare implications of the high-density, caged systems in which chickens are raised, the industry has pivoted to enable consumers to feel good about purchasing free-range and cage-free chicken and eggs.

Layers

Up until the 1990s, Australia produced its own egg-laying chooks. However, because of the genetic advances achieved internationally, the commercial industry in Australia today imports fertile eggs for its layers from overseas. The Hy-Line Brown is from the USA, while from Europe, we import the ISA Brown, Hisex and Lohmann. Grant, a heritage poultry breeder from NSW, thinks this is a real worry. He described to me how he sees the changes in the industry over the decades of his involvement:

> When I was growing up, the vast majority of poultry farms for eggs had purebred sales. And now, the hybrids

> have taken over...ISA Browns and Hy-lines. The pure breeds—the White Leghorns, the Australorps—those old breeds that were being used here were just gupped by the commercial sector. To me, that was a real shame that those breeds that had serviced Australia so well were dumped in favour of overseas breeding lines. It meant that Australia became dependent on European breeders for its egg security. If something happened to the European industry, what sort of ramifications will it have for the Australian industry in time?

The imported hens are prolific layers producing, on average, 340 eggs within the 18 months of what is considered their productive life. Most are then slaughtered, though some are on-sold as rescue chickens, but their prolific laying takes a toll, and health issues, such as reproductive tumours and peritonitis, mean they usually succumb much sooner than heritage birds, who will live ten years or more. Male layer chicks, being unsuitable as meat birds and obviously unable to lay, are identifiable through colour sexing and usually culled at one day old.

As with broilers, there are concerns about the welfare of layers. Stocking densities in small cages curb chickens' capacity to exercise their natural behaviours like wing-flapping, perching, dust-bathing and foraging. This negatively impacts musculoskeletal and metabolic health and can increase stress and aggressive behaviours. This, in turn, results in producers clipping beaks, further limiting chickens from expressing their natural capabilities. In recognition of such issues, conventional layer hen cages will be phased out in Australia by 2036.

While this is a significant win for animal welfare, industry representatives think this is a bad move. They point to challenges with free-range production, including predation from eagles, foxes, snakes and other animals, and the potential for disease transmission through exposure to wild birds. Producing free-range eggs is more expensive because it requires more land, and free-range hens lay around 20% less in winter than hens confined indoors, where temperatures and lighting are kept constant. In the past few years, this has resulted in some periods of egg shortages. Accommodating such challenges is, however, surely a reasonable compromise for higher welfare outcomes.

In being bred for prolific egg production, layers have become less robust. Accordingly, commercial producers institute careful management of light, temperature and humidity, along with significant veterinary interventions to manage disease. This can mean that they do not fare as well in outdoor systems. One agroecological farmer in Queensland has over the years raised both commercial Hy-lines and the heritage utility Australorps. He described his Hy-lines as 'Frankenchickens', who had little 'chookiness' remaining in their character. They were not enthusiastic foragers, and when there was an eagle or another predator nearby, while the Australorps would run to safety, the Hy-lines would 'stare at the sky vaguely' having 'no self-preservation instinct'. He also found that they were aggressive and would attack each other when under pressure, relative to the more placid Australorps.

As with pigs, disease risk has always been high with chickens. They are susceptible to infectious bacterial and viral diseases, metabolic conditions, and parasitic and behavioural afflictions. Most worrying currently is the increasing global prevalence of highly pathogenic strains of avian influenza, otherwise known as bird flu. Recent strains of avian influenza not only threaten domestic poultry, but also have alarming mortality rates in humans. Avian influenza has already killed millions of wild birds and mammals worldwide.

Intensive poultry farms can be a key disease incubator, where overcrowding of genetically similar animals makes for a ripe environment for disease transmission. Biosecurity measures are tight in poultry sheds, yet the list of possible contaminants and breaches is long. Free-range birds are also susceptible to exposure from wildlife. At the time of writing, avian flu outbreaks have been detected in Victoria, NSW and the ACT. Experts suggest such zoonoses will only continue to increase in prevalence over time.

A final concern with the dominance of commercial layers and broilers is that the mindboggling numbers of commercial chickens obscure the fact that breed and bloodline diversity is being lost in spades. An estimated 9% of chicken breeds are already extinct, and a further 30% remain at risk of extinction globally. Conserving breed and bloodline diversity not only mitigates against disease risks, but also ensures the perpetuation of a wide array of traits to meet the needs of future populations.

Ernest and I were discussing this and, along with the loss of so many unique breeds, he worries about the homogenisation of genetics in commercial layers and broilers. He points out that: 'poultry meat is the most consumed meat in the world, and if tomorrow a new virus comes in and wipes out the main ISA Brown egg-laying chooks and the Inghams meat birds, we really need that back-up of rare and different breeds as a safeguard'. The fact that susceptibility to Newcastle disease, as one example, varies considerably between chicken breeds supports Ernest's concerns.

Heritage Chooks

In contrast to the massive sheds of intensive poultry production, Ernest's Araucanas lead a salubrious life on his suburban block. As seen throughout this book, heritage animals tend to be raised on small-scale farms—or, in this case, backyards—with fewer energy, commercial feed, fertiliser, pesticide and herbicide inputs, as well as minimal waste outputs. Heritage animals have been selected over generations to thrive in outdoor systems.

An admirer of regenerative farming approaches, Ernest has turned his quarter-acre block into a rotational grazing utopia. He has a main yard area with three peripheral chicken runs, each with its own pen, which the Araucanas are transferred into on average every three months. At any one time, two of the runs are rested. Then, at different times, the poultry are allowed into the main yard area. All of the runs and pens were built with recycled building materials; in his words 'they are not fancy but very functional'. When he rests a run and pen, he sows grass, bok choy, silverbeet, wheat and other leafy vegetable seeds. Over three or more months, they are left to grow. The trick of the whole system, he says, is to keep green vegetation available all year-round.

Each element of Ernest's system is carefully thought out. He chose the soft feather light Araucanas for their temperament, as he has found the roosters to be calm and minimal crowers, so more feasible to keep given council regulations. (Interestingly, other chook aficionados tell me Araucanas are known to be flighty—'insane' was the specific word one woman used. Different bloodlines within a breed can have different qualities, but it's also certainly the case that perceptions over which breeds are superior are equally varied and fascinating.) Ernest went on to tell me that he believes Araucanas are 'particularly intelligent' and have many additional qualities besides:

> I find that they probably handle the heat better—because we're getting hotter and hotter summers here on the coast. Whereas the bigger birds, they really feel it, particularly if they come from the northern hemisphere...So it's all about survival, egg production, a different type of breed.

The poultry, he said, save him labour as they eat down his vegetation. Not having to mow lawns means he doesn't need a petrol mower. His chooks keep the veggie-eating insects to a minimum. He has a rainwater tank that gravity feeds into a pond that grows watercress that he and the chooks love, and which is home to Australian White Cloud fish who coexist peacefully with frogs, of whom Ernest is also very fond.

Poultry droppings are a wonderful fertiliser, and Ernest places recycled black plastic under his chicken perches late in the afternoon. In the morning, he hoses the droppings off around his lemon tree and veggies. This results in abundant green vegetation, and the more of this he has available for the poultry, the less grain he needs to go out and purchase. He tries to carry out lots of these little savings with the help of his chooks. He sells his Araucana hens' characteristic blue eggs for some income in his retirement, and eats eggs himself, along with any excess roosters.

Albert Bartlett famously quipped that modern agriculture is 'the use of land to convert petroleum into food'. Ernest, like many heritage breeders, is concerned about fossil fuel dependence in intensive poultry production. Commercial egg producers genetically select for hens with the least maternal instincts, as when they go broody, they cease laying. Yet, Ernest appreciates maternal instincts, with the temporary loss of egg production made up for by lower energy costs and environmental impacts, higher hatching rates and the satisfaction of allowing the birds to express a greater spectrum of their natural behaviours.

Consequently, Ernest believes that, even though his heritage birds are slower growing and produce less in the short term, their hardiness, longevity and relatively minimal resource inputs and waste outputs make them a better choice than commercial hybrids in the long term.

Loving Chickens

Far more than a sustainable food source, heritage chickens are valued as animate beings who provide companionship for their breeders. This was particularly pronounced during the Covid pandemic, where on chicken pages on Facebook, people posted a great deal about their chickens' amusing antics. Breeders have repeatedly described to me how their stress dissipates as they spend time feeding, watching and just generally being with their chooks. They develop strong attachments not only to their chosen breeds, but also to particular individuals. The flipside of this love, however, is the grief that comes with the inevitable death of their birds. Heartfelt eulogies are posted frequently in online chicken forums, where people write long paragraphs of text about the qualities and values of a particular chicken. Some go as far as getting tattoos of favourite chooks, while others commission paintings or commemorate them with photos.

There is also a whole language and subculture around chicken keeping. Poultry lovers have a seemingly insatiable desire for more birds, which is encapsulated in the concept of 'chicken maths'. This unique and highly variable approach to counting one's chickens serves to justify additional chicken acquisitions. It is mobilised to assure less galline-inclined spouses, neighbours and councils that the number of chickens owned by any friend of the feathered is fewer than standard numerical conventions would indicate. Some of the rules of chicken maths include:

- Chicks don't count.
- Chooks with names don't count (they're pets, not chooks).
- Birds in moult don't count (with all those feathers missing, they're hardly a complete chook).
- You can count eggs, but not chickens (even the most curmudgeonly spouses/neighbours appreciate egg bounty).
- A flock is happiest and easiest to manage when a ratio of one rooster to around ten hens is observed, consequently, each unit of 11 equals one chicken.
- Bantams (being small in stature) only count as half a chook.
- For those keeping multiple breeds, all those of one breed equal one chicken.
- Similarly, all chooks sharing the same colour constitute a singular chicken.
- Any council's maximum number of chickens is how many chickens you own, minus one, so you can still buy another chicken.

Having so many chickens, however, comes at considerable expense. Some people describe making modest profits or breaking even, but most describe the cost of keeping chickens as more than they return. A common joke is 'how do you become a millionaire by raising chickens?' to which the retort is: 'start out a billionaire'. The time, expense and labour invested in poultry rearing is beautifully captured by Welsh Harlequin duck breeder Eli when describing her plans for the weekend:

> Build pens, build yards, build more pens, build more yards (note to self, stop hatching ducklings…More due on Wednesday). Clean pens. Clean more pens. Rewrite note to stop hatching ducklings. Sell kidney and buy more food for ducks.

Transylvanian Naked Neck Cockerels (Bantams) courtesy of Scott Carter

Heritage Utility Birds

While some raise chickens primarily as pets or for showing, others raise them expressly for food. Tensions between these positions play out in differing views on whether to take a sick chook to the vet. Those in the chooks as pets camp will do what they can afford to keep their pets alive. Those who see them as a utility animal think that treating chickens like a pet or family member is over the top given the cost of replacing the chook is, for the most part, considerably less than seeking treatment. The latter see most issues as either home treatable or likely to lead to death regardless of intervention, with some treatments even prolonging the chicken's suffering for the sake of human emotional attachment rather than the animal's welfare.

For those who love their chooks, but also value their utility, killing excess roosters for meat can be a difficult task. As noted via chicken maths, while it varies based on breed, most consider a ratio of one rooster to around ten hens to be optimal. This is because roosters are highly sexually motivated, so can overwhelm and injure a smaller number of hens. Consequently, given roughly equal numbers of male and female chicks are hatched, there are always excess cockerels. Talking to one breeder, she described taking a life as invariably hard, but felt that this was the appropriate human response. Being desensitised to this, or having no emotional response, would indicate one was a 'psychopath' in her view. She described reading a prayer to her roosters the night before slaughtering, which she conceded was more for her comfort than theirs. However, expressing gratitude for the lives of her animals is an important positioning and practice within her values and is a sentiment shared by many.

The difficulty of killing is seen as balanced by the culinary rewards and satisfaction of producing one's own chicken meat. Of the heritage breeds, there is much dispute about which is the best eating, but Indian game, Dorking, Brahma and Faverolle are some common favourites. Anne, an enthusiastic Faverolle breeder, had the following to say about the gastronomic qualities of this breed:

> I have just had a meal from a homegrown cockerel. It is nothing like a supermarket chicken. This flavour! The meat is generally somewhat darker. It has a silkiness to it. It has a similarity to lamb shank—that stickiness when it is slow cooked. The breast meat is succulent, not the firmness of the bought sample...Some say the time to slaughter is the weekend after the first crow! Yes, that would be about right...I have no idea of the cost to rear, but I suspect that it is more than the $10 of the supermarket chicken. But then, that flavour has no price.

Her comments about the price differential are pertinent. While many consumers react with distress when learning about some of the routine practices in the mass production of broilers, there is a disconnect between this concern and how much people are prepared to pay for more ethically produced chicken. Heritage breed commercial chicken meat producers are as rare as hen's teeth, because economic viability is so difficult to achieve when consumers are so habituated to spending so little on chicken.

In addition, it is increasingly difficult, if not impossible in many regions, for small-scale poultry breeders to access abattoirs. This is owing to the vertical integration of industrial chicken production, where the whole cycle from incubation to slaughter and processing is undertaken in house. For independent processors, the miniscule profit margins per bird, relative to the labour-intensive task of processing, make it a difficult business proposition. Moreover, heritage breeds come in different shapes and sizes, which are more complicated to manage in highly mechanised systems designed around homogenous commercial hybrids. In Victoria, a butcher I interviewed told me there was only one abattoir remaining in the whole of the state that catered to small-scale chicken producers.

Notwithstanding such difficulties, Janek produces meat and eggs from a range of heritage poultry breeds on his organic farm on the lands of the Melukerdee and Lyluequonny people in Tasmania. He has Barnevelder chickens, Swedish Blue ducks, Blue and Black heritage turkeys, Guinea Fowl, pheasants and English game. Janek's family has a long history of raising poultry, first in Poland and then in Australia. His grandfather lost much of his family during World War II. He survived the concentration camps and, for him, poultry was about caring for family through self-sufficiency. Along with keeping poultry and livestock, he produced his own grain, vegetables, fruit, beer and wine, and had seven years worth of preserved food stored at any given time. Of his grandfather, Janek told me:

> He'd lost everything, his entire family, in World War II. He'd been faced with the Nazis and faced with the Russians, the cruelty and the devastation, and the thing of what could happen. He's come out with that mindset that 'I need to supply my needs, and my family's needs, and I need to be able to do that and only rely on myself'.

Janek's grandfather was deeply embedded in the Polish community, with whom he bartered his fowl, but traded with Italians and others, too, 'because all the immigrants were into poultry'. For Janek's grandfather, poultry were not pets; it was always about practical utility. 'What do I need to do to put food on the table?'

Janek says he was around five years old when his grandfather and mother showed him how to kill and process chickens. There was no time for squeamishness; he remembers learning how to use the axe, and his mum cooking a beautiful roast chicken. He believes eating a hen at the end of her laying life is the greatest act of respect you can show. Janek told me he does everything he can to provide his poultry with a good life. They forage not only on pasture, but also in bushland. He thinks confining fowl in sheds is cruel, and he loves watching his birds express their full range of behaviours. Like Ernest, Janek also encourages maternal instincts. Broody hens mean he's less reliant on electricity for incubation. I ask about hatching rate, to which he responds:

> When a mother hatches, we rarely lose chicks, and our hatch rate, our fertility rate, because we've got a program that is heavily out-crossed, and so our birds are not just very healthy, but they're very fertile...But we also have a very high survival rate, which is fantastic because you put other eggs in an incubator. I've heard of hatch rates as low as 30, 40% up to 60, 70, 80%. But often we're getting 100% hatch rates and 100% survival rates. You couldn't ask for more than that.

Janek uses no pesticides, herbicides or antibiotics. If a bird is sick, he culls it, which he believes ensures he breeds only the healthiest, most resilient animals. He sees poultry as an integral part of any agroecosystem. To this end, he calls his Barnevelders 'triple purpose', as they are great to use in permaculture systems, eating bugs and fertilising the land, alongside being good layers and good eating. I asked if predation was a problem and he has had some issues with quolls killing poultry, but rather than killing the quolls, he has moved the fowl. He believes his system respects the natural environment and its local native species, and that agriculture must adapt to and work with local conditions.

Janek's devotion to his poultry has, at various stages of his life, come at a cost. His partner left him for a time because, after the death of two family members who were partners in the family poultry business, not only was he coping with grief, but he also had a steep learning curve to take control of the operation, resulting in 80-hour working weeks. His partner issued an ultimatum: if he wanted to continue the relationship, he had to get rid of the poultry. He recounted his reply: '"Look, I can't do that. It's family, it's part of my family heritage. As much as I love you"—and we were talking about marriage—"you're asking me to do something that, if I did it, I would resent you for the rest of my life."' As she was leaving, he said she walked out to the paddock and, in a poignant moment, told his beloved Swedish Blue ducks that they had won. Over the year that followed, however, she came to feel that her request had been unreasonable and their relationship resumed. Janek's loyalty and commitment to his poultry is evident here, even at the risk of great personal loss.

Showing Chooks

Janek's grandfather was greatly respected for the quality of his poultry, but he was never involved in showing his birds, despite this being a popular pastime and a key social forum in poultry communities. There is, at times, tension between the utility and showing camps. The *Australian Poultry Standards* is the bible for exhibition poultry breeders. At shows, poultry are judged primarily on their physical qualities, and how they measure up against these breed standards. Since showing has become such a central part of heritage breed chicken culture, some are concerned that the utility qualities of heritage breeds have diminished due to selective breeding for appearance over laying or meat-providing capacities. As one breeder described: 'The meat and egg production, once prized in these old breeds, has been sacrificed for ribbons at

Red Junglefowl Pair courtesy of Scott Carter

poultry shows.' He went on to tell me how his goals were to 'breed a robust, personable, beautiful bird, which continues into old age in great health without ongoing vet bills, while still keeping the family in eggs, and for the surplus boys to grace the table'.

Many others see poultry shows as integral to efforts to conserve heritage breeds. Grant has been showing fowl since the early 1980s. He breeds Sussex (white and speckled), Campines (gold and silver), White Leghorns, Australorps, Rhode Island Reds and Barred Plymouth Rocks. In 2006, he and some like-minded breeders established a rare breeds poultry club when they started to notice a decline in numbers and popularity of several heritage breeds. For him, showing and utility need not be mutually exclusive. His interest has always been the productive attributes of fowl. He is passionate about promoting and preserving the old breeds, which he enacts through selling fertile eggs at low prices, between $15 and $25 a dozen. He explained his rationale as follows:

> If you want to promote them, and get good numbers of them out there, you've got to make it so that the kids that are interested in rare breeds can afford to buy a set of eggs with their pocket money. I mean, people who charge $60, $65, $70 a dozen, that's putting it out of the realm of kids. You've got to get young people interested in things if there's going to be any future. In 20 years time, when I die, there's nobody coming on to replace me. What's it all been for? Just for my own gratification?

As part of his conservation efforts, he will take back any unwanted cockerels. This works well for those who can't bring themselves to kill them, and for Grant, as it increases the pool of roosters from which he can choose to breed:

> I can use them in the breeding pen if I need to. If I don't, I'll eat them. It doubles the amount of males that I can rear and keep myself. It's very nice to have someone that you've sold eggs to bring you back a beautifully grown male bird that you haven't had to feed!

As we saw in Chapter 3, among many in the heritage breed community, value is seen in terms of conservation and promotion of the breeds above purely economic factors.

Since settlement, Australians have highly valued chickens, turkeys, ducks, geese and game fowl, enjoying their eggs and occasionally meat in return for feed and care. Over the past half-century, however, the scaling up, intensification and selective breeding of chickens, particularly, has dramatically accelerated meat and egg production at a considerable cost to animal welfare. Heritage breed chook enthusiasts worry about the long-term health of the species within the current paradigm. They are also concerned about the loss of breed diversity across all domestic poultry species. For many breeders, while they appreciate their poultry as a sustainable food source, they also treasure their birds as quirky pets. Many report simply enjoying spending time with their fowl, while others relish the challenge of breeding to the highest standard, and the sense of community gained through showing their various breeds. As with other heritage livestock species, it is the breeders love of their chosen breeds that is serving as a safeguard for agrobiodiversity conservation into the future.

Concluding Thoughts

The entanglements of humans and domestic animals at any scale and in any system are complex. Since mid-last century, the livestock industry's scaling up and intensification has come at a cost to the environment and animal welfare. The construction of animals as commodities has ruptured the relations of reciprocity that have long benefited both humans and animals. In Australia, manifestations of climate change in the form of the catastrophic droughts, fires and floods the country is experiencing so frequently have made the urgency of learning to live less destructively, and in ways respectful to all life forms, clear. Agriculture can either exacerbate the problems we face, or it can be a critical part of the solution.

Animal consumption is deeply entrenched in dietary practices and the nation's economy. Consequently, supporting ecologically sustainable and higher animal welfare practices is critical. Heritage breed cattle, sheep, pigs and poultry embody agrobiodiversity, and these breeds have demonstrated over time that they can survive and thrive in diverse and marginal environmental conditions. In the climate change era, heritage breeds and the systems in which they're raised can serve as both an ode to the past and a safeguard for an uncertain future.

The risk of extinction many heritage breeds face heightens the stakes of interspecies relationships. After a long conversation with Sue in the early stages of the project, she emailed her reflections on breeders of heritage livestock: 'You will find the one thread that binds them all together is passion for their chosen breed, and they have it in spades. I think that is the one thing that may save our rare and heritage livestock.' The passion Sue speaks of, the intensity of emotion, is usually expressed by farmers quite simply as love. And love is a powerful tool for conservation.

Sue went on to observe that 'breeders will never be rich financially. But there is a different kind of wealth that comes from knowing you have done something to help a rare breed to continue for a future generation, and that is a lot more satisfying.' Sue concluded her warm, eloquent message by suggesting I was now a part of this tapestry. I would like to end by extending her invitation to readers to support small-scale farmers in their efforts to produce food in ways that support ecological integrity, animal welfare, human flourishing and the maintenance of breed diversity into the future.

Contrary to much of the discourse we hear from agribusiness, the FAO confirms that approximately 80% of the world's food is produced on family farms. Moreover, family farms tend to be ecologically sustainable and foster more vibrant rural communities, while being home to greater wildlife variety and agrobiodiversity. Accordingly, supporting small-scale farmers is key. We'll give the last word to Janet, a passionate advocate of heritage breeds, who spells it out:

Until we get good governance, the best we can do—and it's a lot—is to shop ethically. Buy local. Don't buy foreign owned. Buy 100% Australian owned, 100% Australian made. If there is no choice, go without or get something else. Check who really owns brands. Avoid intensive farming. Tell supermarkets where to put their cheap milk. Free-range chicken now outnumbers misery chicken in most supermarkets—due to consumer demand. We can do the same for pork, beef, lamb, dairy. A small local cheese is much better than a big cheap one from a foreign-owned dairy where cows are numbers. Quality, not quantity. Go to farmers' markets. Buy to give our farmers a living, and for our health—you will probably be buying the produce of a rare breed. Go, you good thing, you!

Chapter Notes and Further Reading

Introduction

Much of the data in this book comes from my conversations and time spent with farmers during my fieldwork across Australia from 2020 to 2024. Depending on the farmer's wishes, either their real name or a pseudonym havsbeen used.

The book has also been shaped by a large body of interdisciplinary research and conservation work spanning the agricultural and social sciences, advocacy groups and industry. International efforts to monitor and protect breed diversity in livestock are spearheaded by the FAO's Commission on Genetic Resources for Food and Agriculture. They produce regular comprehensive reports to document the state of livestock diversity globally. I draw particularly on their 2015 publication *The Second Report on the State of the World's Animal Genetic Resources for Food and Agriculture* (edited by B. D. Scherf and D. Pilling) and their 2019 report *The State of the World's Biodiversity for Food and Agriculture* (edited by J. Bélanger and D. Pilling).

In Australia, the monitoring and conservation of livestock breeds is left in the hands of non-governmental organisations. The Rare Breeds Trust of Australia undertakes much of this work, but is run entirely by volunteers, with the sole source of income being membership fees and fundraising. In 2021, I was invited to join the RBTA Board, and I have served as a director and secretary since this time. Successive Australian governments have shown little interest in livestock breed conservation, so we undertake the national livestock breed censuses for the FAO. We promote rare breeds via our website and a very active Facebook page with over 10,000 members. Our current focus is on establishing a germplasm bank to store the semen, and eventually embryos, of endangered livestock breeds. My role with the RBTA has allowed me to learn a great deal about the behind-the-scenes work volunteers undertake in pursuit of the preservation of breed diversity.

Significant research in biological and agricultural sciences documents and analyses the issues explored in this book. Stephen Hall's *Livestock Biodiversity: Genetic Resources for the Farming of the Future* (Blackwell, 2004) is a particularly useful and thorough examination of breed diversity, its loss and implications. The various definitions of the breed concept I offer in the Introduction are drawn from this book particularly. For a succinct synopsis of the issues, I also recommend the 2008 article by Taberlet et al. entitled 'Are cattle, sheep, and goats endangered species?' which is published in *Molecular Ecology* (vol. 17, pp. 275–84).

In contrast to the extensive agricultural sciences literature, attention to livestock diversity is more fledgling in the social sciences and humanities. Yet, this is important work given breeds are determined as much by cultural values and practices as environmental processes. This is amply illustrated in Rebecca Wood's history of the development and dispersal of various breeds in *The Herds Shot Around the World: Native Breeds and the British Empire, 1800–1900* (University of North Carolina Press, 2017). How we categorise and make sense of breeds is also insightfully explored by the historian Harriet Ritvo and geographer Lewis Holloway, in numerous works.

On love, my ideas have been influenced by ruminations offered in various works by both Lauren Berlant and Michael Hardt, and particularly the work of anthropologist Deborah Bird Rose. In *Wild Dog Dreaming: Love and Extinction* (University of Virginia Press, 2011), she makes a compelling case for the power of love in conservation. She touches on First Nations communities' engagements with cattle and other introduced species, alongside human–animal relations broadly, including the need to reconfigure interspecies connections in the age of extinction.

Bird Rose's ideas are supported by research on the positive implications of loving for humans and animals in agricultural contexts. The work of Xavier Boivin is particularly interesting in this respect. In his 2003 'Report on Stockmanship and Farm Animal Welfare' published in *Animal Welfare* (vol. 12, pp. 479–92), coauthored with Joop Lensink, Céline Tallet and Isabelle Veissier, the health, welfare and performance gains for animals treated with positive handling practices are evidenced.

Throughout the book, in recognition that all farming described takes place on unceded lands, I include reference to First Nations Country names. This information was either provided by farmers or the AIATSIS website: https://aiatsis.gov.au/explore/map-indigenous-australia.

Chapter 1: Cattle

Anthropologist Cristina Grasseni's work exploring heritage Alpine cattle breeds in Switzerland offers a fascinating and detailed account of the value attributed to traditional breeds in the high country. Her writing forms part of the broader multispecies ethnography movement, which constitutes an important shift in anthropology towards decentring the human. Radhika Govindrajan is another influential contributor to this literature, and her wonderful book *Animal Intimacies: Interspecies Relatedness in India's Central Himalayas* (University of Chicago Press, 2018) explores several categories of interspecies relatedness. How people in Uttarakhand relate to local heritage breed cattle versus imported commercial breeds powerfully illustrates the symbolic and religious significance attributed to traditional breeds. She also writes evocatively of the role of love in human–animal relations.

For further reading on agroecology and the problems with the dominant food system, Peter Rosset and Miguel Altieri's *Agroecology: Science and Politics* (Practical Action Publishing, 2017) is a great starting point. For an Australian perspective on regenerative agriculture, see Charles Massy's influential tome, *Call of the Reed Warbler* (University of Queensland Press, 2017). On questions of the impacts of livestock especially, both positive and negative, Henry Janzen offers an excellent synopsis in his 2011 article 'What place for livestock on a re-greening earth?' in *Animal Feed Science and Technology* (vols. 166–167, pp. 783–96).

The relationship between breed and genetic diversity is a complex one, with important implications for conservation. Felius et al. explain this as part of their overview of the history and development of cattle breeds in their article

'Conservation of cattle genetic resources: The role of breeds' published in 2015 in the *Journal of Agricultural Science* (vol. 153, pp. 152–62).

The jaw-dropping state of Holstein cattle genetics is laid bare in the work of Xiang-Peng et al. in their 2015 article 'A limited number of Y chromosome lineages is present in North American Holsteins' published in the *Journal of Dairy Science* (vol. 98, pp. 2738–45). On Tarantaise cattle, whose susceptibility to heat stress was found to be less than that of the hyper-productive Holsteins, see the 2016 article by Bellagi et al., 'The interest of a mountain dairy cow breed to cope with Mediterranean summer heat stress' in I. Casasús and G. Lombardi (editors), *Mountain Pastures and Livestock Farming Facing Uncertainty: Environmental, Technical and Socio-economic Challenge* (CIHEAM, pp. 143–7).

Given their popularity, there is a vast literature on Angus cattle. The cattle industry provides statistics on breed numbers on websites such as beefcentral.com, while an overview of disease issues can be found in the article by Whitlock et al., 'Heritable congenital defects' in *Bovine Reproduction* (edited by Richard M. Hopper; John Wiley, 2021).

On human–bovine relations, the article indicating that dairy cows with names had higher productivity was C. Bertenshaw and P. Rowlinson's 2009 article, 'Exploring stock managers' perceptions of the human–animal relationship on dairy farms and an association with milk production' published in *Anthrozoös* (vol. 22, pp. 59–690). For those interested in human–livestock relations, I have argued that heritage breed farmers see animals as family, and explore the complexities of loving, kinning and killing animals in the chapter 'Blood ties: kinning and killing on Australian heritage breed farms' in *Nurturing Alternative Futures:*

Living with Diversity in a More-than-Human World (edited by Muhammad A. Kavesh and Natasha Fijn; Routledge, 2023).

Chapter 2: Sheep

There are several fascinating accounts of the introduction of sheep and other livestock species to Australia. Ian Parsonson's *The Australian Ark: A History of Domesticated Animals in Australia* (CSIRO Publishing, 1998) is among the most comprehensive, while other classics include Ted Henzell's *Australian Agriculture: Its History and Challenges* (CSIRO Publishing, 2007) and Geoffrey Blainey's *A Land Half Won* (MacMillan, 1980).

More recent histories focus on the dispossession of First Nations people and the role livestock has played in the nation's violent colonisation. Cameron Muir's *The Broken Promise of Agricultural Progress: An Environmental History* (Routledge, 2014) is essential reading in this regard, and is complemented by Christopher Mayes' *Unsettling Food Politics: Agriculture, Dispossession and Sovereignty in Australia* (Roman & Littlefield, 2018).

From his unique position as both a farmer and scholar, Charles Massy is today one of the most influential commentators on the sheep industry, specifically, and Australian agriculture more broadly. He casts a critical eye on the wool industry's demise in his book *Breaking the Sheep's Back: The Shocking True Story of the Decline and Fall of the Australian Wool Industry* (University of Queensland Press, 2021).

Every sheep farmer's worst fear played out in the foot-and-mouth disease epidemic in the UK in 2001. For a fascinating examination of what the disease reveals about human-ovine relationships, I recommend the 2005 article by Convery et al., 'Death in the wrong place? Emotional geographies of the UK 2001 foot and mouth disease epidemic' published in the *Journal of Rural Studies* (vol. 21, pp. 99–109). Also focusing on this epidemic is Harriet Ritvo's excellent 'Counting sheep in the English Lake District' published in Dorothee Brantz' (editor) *Beastly Natures: Animals, Humans, and the Study of History* (University of Virginia Press, 2010). Staying in this region, in a more popular vein, James Rebank's delightful recounting of his intergenerational farming family's commitment to the shepherding life, *The Shepherd's Life: A Tale of the Lake District* (Penguin Books, 2015), is the perfect rainy day read. Sheep farmers in Australia retain a close interest in and relations with sheep farming in the UK, and there is much common ground, notwithstanding the very different environments in which sheep are raised.

Chapter 3: Pigs

Australians have an ambivalent relationship to pigs. On the one hand, pigs affectionately feature in popular culture (think *Babe* and *Charlotte's Web*) and pork is a favoured meat. On the other, in their feral manifestation, pigs are subject to intense loathing. On the latter, a fascinating portal to the past is available in EM Pullar's 1953 article 'The wild (feral) pigs of Australia: their origin, distribution and economic importance' published in the *Memoirs of Museum Victoria* (vol. 18, pp. 7–23), which offers a useful history of feral pigs in Australia. For a more contemporary analysis, see Paul Keil's 2023 article, 'Unmaking the feral: the shifting relationships between domestic-wild pigs and settler Australians' in *Environmental Humanities* (vol. 15, pp. 19–38).

From the wilds to the factory farm, Alex Blanchette's deep-immersion ethnographic research undertaken in intensive piggeries in the USA comes to life in his nuanced and excellent book, *Porkopolis: American Animality, Standardized Life, and the Factory Farm* (Duke University Press, 2020). While there are many commonalities across the increasingly globalised livestock production sector, much can be learned about the specificities of the Australian context from the industry itself: https://australianpork.com.au/.

Critical of this industry, and the animal health and welfare challenges posed by intensive confinement, are grassroots organisations such as AFSA which, by its own description, is 'a farmer-led civil society organisation of people working together towards socially-just and ecologically-sound food and agriculture systems that foster the democratic participation of Indigenous Peoples, smallholders, and local communities in decision making processes' (https://afsa.org.au/). I have been a member of AFSA since 2017 and served for a time on the National Committee. AFSA seeks radical transformation of the food system and advocates for an agroecological transition. It works to support pig (and other) farmers raising animals in pastured systems that allow them to express their full range of behaviours.

AFSA also helps farmers tell their stories so that consumers can better understand where their food comes from. These stories can be mobilised to encourage consumers to aid heritage breed conservation efforts. While in its infancy in Australia, where breeds are not widely recognised by consumers, anthropologist Brad Weiss' *Real Pigs: Shifting Values in the Field of Local Pork* (Duke, 2016) shows how the desire for authenticity, and ethically and sustainably produced food, is driving growth in the heritage pig economy in the USA, which is excellent news for heritage breed conservation there. As breeders often say, 'we have to eat them to keep them'.

Chapter 4: Poultry

Recent research has changed our understandings of chicken domestication and distribution. Joris Peters and collaborators undertook extensive analysis of archaeological data to determine that chicken domestication and distribution occurred considerably later than previously understood. Their findings are described in their 2022 article, 'The biocultural origins and dispersal of domestic chickens' in *PNAS* (vol. 119, no. 24). A consortium of researchers led by Julia Best has also enlivened our understanding of human–galline relations through their data's revelation that chickens were kept as exotica for centuries before being considered food in Europe. Findings appear in their 2022 article, 'Redefining the timing and circumstances of the chicken's introduction to Europe and North-West Africa' published in *Antiquity* (vol. 98, pp. 868–82).

The chicken industry has a strong web presence, where lots of information and statistics on contemporary chicken production and consumption can be found. On eggs, see https://www.australianeggs.org.au/egg-industry, and on meat chickens, see https://chicken.org.au/. For a more critical assessment of these practices, Annie Potts' *Chicken* offers an excellent overview (Reaktion Books, 2012). Drilling into the broiler industry in the USA, and making a compelling argument about the work we make chicken bodies do, is Les Beldo's 2017 article, 'Metabolic labor: broiler chickens and the exploitation of vitality' published in *Environmental Humanities* (vol. 9, pp. 108–28).

For historical insights into the changing dietary patterns in Australia around chicken consumption, Jane Dixon's *The Changing Chicken: Chooks, Cooks and Culinary Adventures* (UNSW Press, 2002) is illuminating. Historians have much of interest to say about chickens, and also worth perusing is Catherine Oliver's take on the complexity of a species that has such vast numbers, yet such vastly diminishing diversity. Her analysis can be found in her 2022 article, 'The opposite of extinction' published in *Environment and History* (vol. 28, pp. 197–202).

On the variable susceptibility of certain chicken breeds to a nasty galline disease, see Megan A. Schilling and colleagues' 2019 article, 'Conserved, breed-dependent, and subline-dependent innate immune responses of Fayoumi and Leghorn chicken embryos to Newcastle disease virus infection' in *Nature: Scientific Reports* (vol. 9, pp. 1–10).

Finally, Albert Bartlett's much-cited definition of modern agriculture as 'the use of land to convert petroleum into food' can be found in his talk entitled 'Reflections in 1998 on the twentieth anniversary of the paper, "Forgotten fundamentals of the energy crisis"'.

Acknowledgments

I would like to begin by acknowledging that this book was written on the unceded lands of the Whadjuk Noongar people, who have shared stories here for millennia. I acknowledge their strength and resilience over the past two centuries of colonialism, and the complex role of livestock in the history of First Nations dispossession across the country.

This book owes its existence to the dozens of farmers who shared their fascinating stories with such generosity and enthusiasm. Their passion is contagious, and my only regret is that I couldn't include more people's stories in this short book. My gratitude to each and every one of you, who work so hard to put food on our tables, while nurturing the land and the wonderful diversity it contains.

This research was funded by the Australian Government through the Australian Research Council's Discovery Project scheme (project number DE200100595). This funding also supported Tammi Jonas and Domenico Volpicella to undertake their PhDs within the project. It was such a pleasure thinking and learning with them, and co-supervisors Greg Acciaioli, Heather Bray and Katie Glaskin, over the past four years. I'd also like to thank Lauri Turner, who contributed to this project through six months of wonderful work as my research assistant.

I'm grateful to the Vignettes series editors, Tony Hughes D'Aeth and Sarah Collins, for encouragement and feedback throughout the process. For reading and commenting on chapter drafts, and sharing your expertise so unreservedly, my thanks go to Judy Barnet, Katy Brown, Donald Cochrane, Sue Curliss, Sarah Czerny, Henry Dunn, Peter Gelmi, Brenton Heazlewood, Kyla Howard, Tammi Jonas, Jan Kleynhans, Susan Locke, Vadim Pantall, Kim and Geoff Rowlatt, Anne Sim, Bella St Claire,

Jill Weaver, Murray Williams, and my number one Copy-Editor-in-Chief, Rosemary Gressier, aka Mum, who read countless drafts, thank you.

For the beautiful images, I'd like to thank Katy Brown, Scott Carter, Sue Curliss, Henry Dunn, Tammi Jonas, and Erica Smith, whose Highlands are also the models for the wonderful photo by Kym Papworth.

For tolerating the endless animal talk over the past few years, I'm so grateful to Abbie, Adam, Alex, Ben, Em, Fel, Hannah, Harvey, Izzy, Jen, Justine, Lara, Matt, Neil, Rita, Saam, Taz, Tobi and Triinu. Finally, special thanks to Mum, Alicia and Nacho, and DelnShano, Tahnee and Michael for keeping Abs entertained during fieldwork and writing frenzies.